Contemporary Irish Studies
Series Editor Peter Shirlow (School of Policy Studies,
University of Ulster, Jordanstown)

Also available

Colin Coulter
Contemporary Northern Irish Society
An Introduction
[Print on demand only]

Graham Ellison and Jim Smyth
The Crowned Harp
Policing Northern Ireland

Paul Hainsworth (ed.)
Divided Society
Ethnic Minorities and Racism in Northern Ireland

Patrick Hayes and Jim Campbell
Bloody Sunday
Trauma, Pain & Politics

Peadar Kirby, Luke Gibbons and Michael Cronin (eds)
Reinventing Ireland
Culture, Society and the Global Economy

Jim Mac Laughlin
Imagining Ireland
The Contested Terrains of Irish Nation Building

Denis O'Hearn
Inside the Celtic Tiger
The Irish Economy and the Asian Model
[Print on demand only]

Gerry Smyth
Decolonisation and Criticism
The Construction of Irish Literature
[Print on demand only]

Gerry Smyth
The Novel and the Nation
Studies in the New Irish Fiction
[Print on demand only]

Belfast

Segregation, Violence and the City

PETER SHIRLOW
and
BRENDAN MURTAGH

Pluto Press

LONDON • DUBLIN • ANN ARBOR, MI

First published 2006 by Pluto Press
345 Archway Road, London N6 5AA
and 839 Greene Street, Ann Arbor, MI 48106

Distributed in the Republic of Ireland and Northern Ireland by
Gill & Macmillan Distribution, Hume Avenue, Park West, Dublin 12.
Tel: + 353 1 500 9599. e-mail: sales@gillmacmillan.ie

www.plutobooks.com

British Library Cataloguing in Publication Data
A catalogue record for this book is available from the British Library

ISBN   0 7453 2481 9 hardback
ISBN   0 7453 2480 0 paperback

Library of Congress Cataloging in Publication Data applied for

10   9   8   7   6   5   4   3   2   1

Designed and produced for Pluto Press by
Chase Publishing Services Ltd, Fortescue, Sidmouth, EX10 9QG, England
Typeset from disk by Stanford DTP Services, Northampton, England
Printed and bound in the European Union by
Antony Rowe Ltd, Chippenham and Eastbourne, England

Contents

Tables, Figures and Maps

FIGURES

MAPS

Acknowledgements

We would like to thank our respective colleagues in the School of Environmental Sciences (University of Ulster) and the School of Environmental Planning (Queen's University Belfast). In particular we thank those who have aided us with a series of insights and recommendations as well as their genuine interest in what we were aiming to achieve: Brian Graham, Jon Tonge, Kieran McEvoy, Dawn Purvis, Michael Murray, Mickey Liggett, Rab McCallum, Billy Mitchell, Chris O'Halloran, Paul Donnelly, Rachel Monaghan, Paul O'Neil, Victor Mesev, Andy Wood, Mike Raco, Nick Phelps, Martin Jones, Rachel Pain, Tommy Quigley, Tom Roberts, Tim Smyth, Adrian Geulke, Jim Smyth, Mike Morrissey, Mike Poole, Shirley Morrow, Dave Eastwood, Bill Neill, John O'Farrell, Ann Mallon, Paul Moss, Ed Cairns, Alan Bairner, Ken Reid, Jim McAuley, Claire Mitchell, Wayne Foord, Charlotte Cox, Joe McGrath, Henry McDonald and Jean Brown.

We would also like to thank the editors of *Regional Studies*, *Urban Studies* and *Ethnopolitics*, who published previous work upon which many of the arguments made within this book are based. In particular, we wish to highlight our appreciation to them with regard to the reproduction of some of the maps, data and other material. Without such support it is always difficult to try out novel ideas and to have the opportunity to work on them. Special thanks to Vincent Gribbin, Dennis McCoy and Stephen Donnelly from OFM/DFM–Equality Unit for sponsoring our various projects on segregation, as well as staff at the Northern Ireland Housing Executive, in particular, Joe Frey. Community groups throughout Belfast have also provided support and encouragement, as have each of the political parties.

Special thanks to the thousands of people who have completed surveys over the years and who have also provided insights from the doorstep and in the parlour. Those who gave us interview material are also to be praised. Many emotive and heart-rending events were recounted by these people. We hope that this book will help create a wider appreciation of the difficult situation that many people in Belfast endure on a daily basis. A final thanks to Oonagh and Eilish.

Introduction

The poet W.B. Yeats's famous line 'for peace comes dropping slow' was long a valid interpretation of the sloth-like pursuit of conflict transformation in Northern Ireland. In 1994 when the first paramilitary ceasefires were called there was a sense that the 'weight of the world' had been removed from the shoulders of the people of Northern Ireland. There was a genuine sense of relief as people viewed the future with a sense of optimism and hopefulness. Eleven years later the Provisional Irish Republican Army (PIRA) decommissioned in a period in which that sense of hope had slowly shifted towards grimace, foreboding and denial regarding the political progress that had been made. During that period of peace-building rioting, petrol bombing and violence had returned in a manner that suggested that the possibility of meaningful civil and political change had been neutered.

In reality, much had changed. There was a new and far from aggressive relationship between the two states and the experiment in devolution had shown signs of a capacity to share power. But on other levels it appeared as though the division between groups remained and, if anything, as argued within this book, the capacity to engender ethno-sectarian atavism has increased. But why would that be the case? How could a peace process lead to a decline in a middle ground within Northern Irish society? Why did peace-building fail to challenge the nature and practice of segregated living?

We aim to answer such questions by highlighting how the meaning of segregation undermines the capacity to shift society onwards. Place is crucial in the reproduction of violence and is as important as ideology itself within conflictual societies. Segregated places are the sites of social exclusion, fatalism and economic truncation. They are also the places within which resistance against economic, social and cultural 'others' is commonplace. The merging of people with shared visions and beliefs, whether imagined or real, into segregated places provides the capacity to fortify togetherness and identity. Ultimately, segregated places are important because they contain the power to subvert and transcend conventional practices and beliefs. Segregated places are a reminder of how conventional society has failed to deliver equality and support for the wider notion of the democratic nation.

1

Belfast's segregated places are important because they are the sites within which the most passionate renditions of difference are played out and managed. In simple terms Belfast has not become a city about which there is no longer much to write. Instead, owing to highly observable forms of spatial division it remains as unfamiliar when we compare it with other cities. It is one of those places within which history, identity and culture are played out, in a dramatic and public manner, through the instruments of threat, menace, violence and deviance. Change, as evidenced by regeneration and additional examples of conspicuous consumption, masks the enduring reality of ethno-sectarian division, poverty and cultural and political dissatisfaction. Belfast is far from being the post-conflictual city that is dreamed of by planners, investors and wearied British and Irish politicians.

Like other cities that have endured the shock of violence and fear, it is a place within which the realities of failure, within the modernisation project, are manifest. Public discourse within Belfast is driven by poverty, sectarianism, racism, myth and cultural hyperbole. These are all facets of city life that can be identified elsewhere. But rarely are such deep-seated problems the stimuli for armed violence and a stubborn refusal to move beyond the concept of national loyalty. If anything, divided cities are places of passion, commitment and unwavering belief. Such passions may be misdirected but they are formed through a reaction to conflict and the reproduction of group differences. Belfast is important not as an economic or civic centre but as a site of conflict and place-centred divisions.

As argued within this book, the failure to resolve ethno-sectarian tensions at ground level and within the most marginalised and deprived communities in Belfast is indicative of a continual refusal to tackle the nature of underlying tensions and realities. There appears to be a common tendency in societies emerging from conflict to present the future as utopian, shared and equal. Such naivety undermines the potential for history, victimhood and new forms of violence to reappoint the desire among most to remain attached to separate ideas, beliefs and practices. In the pages that follow, the presentation of Belfast as a normalising place is challenged, as the reader is reminded that inequality both in material and cultural terms remains as an ever-present force that shapes the reality of wider divisions.

Challenging the myth of a rational and shared vision for societies emerging from conflict is crucial in reminding us of the burdens created by violence. The arguments made within this book are part of a broader debate that challenges the fiction of the United Kingdom as a multicultural society. Denying the existence of multiculturalism is not simply about recognising that rioting that has occurred in places such as Bradford and Harmondsworth. It is about highlighting how people live their lives in segregated communities within which the despair of poverty and social dislocation is increasingly finding a voice in cultural, racial and sectarian affiliations. This book argues that identity matters, but in doing so it explores what identity means and, more importantly, how it is influenced by geographies of ethno-sectarian separation. The key aim of any analysis of ethno-sectarian difference must be not to describe what it is, but to determine how and why separation is employed as the *raison d'être* of political, cultural and social identification.

Lest we forget, the sectarianisation of society in the UK is encouraged by a government intent upon the promotion of faith schools and legislation centred upon religious hatred. The same government subjects the British labour market to the unfettered forces of globalisation but fails to acknowledge the causal link between that process of economic change and the ghettoisation of life along racial lines.

NORTHERN IRELAND: A PLACE APART

With regard to Northern Ireland, it was assumed, by the two states and numerous commentators, that the significant decline in the numbers being killed was a sign of the beginning of the end of conflict. Yet, as shown within this book, the nature of violence has shifted away from paramilitary and state assaults towards a more sectarianised and repetitive violence of interface rioting and attacks upon the symbols of tradition such as Orange Halls, GAA property and churches. We argue that the growth in new and more explicitly sectarian forms of violence has polarised an already divided society. We acknowledge this as being primarily driven through a political process that rewards resource competition between the dominant political parties. Politics in Northern Ireland has been garnered by the capacity to win concessions from the British and Irish states. The violence that has taken place with regard to the right to march, most commonly with reference to the Orange Order, also highlights the reappearance of a form of territorial contention that upholds and encourages ethno-sectarian hostility. This, as well as violence around housing, the location of schools and other territorial conflicts between segregated places, highlights the paucity of a peace process that camouflages the reality that sectarianised allegiances have remained.

As illustrated within this book, fear caused by several decades of violence has undermined the capacity of most within highly segregated places to undertake normal mobility practices within the city of Belfast. However, as shown in Chapter 4, the fear that undermines contact between the two main traditions is augmented by threats from within the host community. The voiceless, or those who wish to move beyond ethno-sectarianism, are studied within this volume in an effort to illustrate the various dimensions of repressiveness within segregated communities. Previous studies of segregation have failed to explore the nature of heterogeneity within segregated places and how those with the most passionate and vociferous attachment to place undermine the ability of 'others' to promote secular and pluralist actions.

Northern Ireland remains as a place apart within a UK and an Irish context. Talk is now not of a divided Ireland but of a divided place that lies on the British side of the Irish border. The policy of Ulsterisation, adopted by the British state during the 1970s, shifted the analysis away from Ireland's problem to a problem in Northern Ireland. Ulsterisation slowly parochialised an already parochial problem. Within this process of political detachment Anglo-Irish dialogue sought to provide a limited role for the southern state in Northern Irish affairs. By the mid-1990s the Republic of Ireland had ditched the myth of a 'nation once again' but was content to have some role to play for 'aul' times' sake, if nothing else. The onset of devolution furthered the provincial dimension and by the end of the twentieth century the chimera of a non-state involvement in conflict in Ireland was, within many quarters, fully fledged.

The end of the twentieth century witnessed a more benign politics of state-building in Ireland, as both main players sought the 'loftier' politics of post-nationalism and the combined forces of Europeanisation and globalisation. However, in Northern Ireland, a place not protected from the new economy and liberal lifestyles, the dominance of an ethno-sectarian logic remained. Here, the ironies and complexities of Anglo-Irish history endured as the heat of political discourse rose despite the onset of paramilitary ceasefires and devolution. Northern Ireland, within the British–Irish context, remained 'a different place' within which peace had dropped slowly (Tonge, 2004).

STUDYING THE CITY

In this book we present a frank and robust examination of a city that has 'moved on', but in a manner that suggests that change has been uneven and at times unjust. In developing ideas of citizenship, the account presented aims to provide a critical assessment of what has failed and what will continue to fail within a contested and spatially divided society. Essentially, we aim to further the ideas of democracy, accountability and critical realism within the context of challenging the ethno-sectarian imagination (Archer, 1995; Bhasker, 1998; Du Gay, 1997). Our analysis attempts not merely to account for fractured identities but to explain the material reality of the divided place that is Belfast. The spirit of the debate contained within this book seeks ways in which to alter the negative nature of weak ties between communities, politicians and policy-makers (Bradley, 1996; Granovetter, 1973; Keat, 2000). In the pursuit of change we conclude that exclusive notions of and devotion to unquestioned group-based identities undermine the pursuit and existence of more pluralist and secularist lifestyles. There is no denying that such group-based identities will endure but, as contended within this book, the Belfast Agreement copper-fastened the importance of groups and in so doing denied not only the heterogeneous nature of Northern Irish society but also the assembly of diverse voices within groups. Political change furthered the exclusive against the exceptional and undermined the rationale of alternative political identities.

We explore the central reasons why conflict between communities remains a dominant feature of living in Belfast. Northern Ireland's administrative centre is also the capital of violence, fear and intercommunity discord. A fundamental reason for the ongoing 'privilege' of being spread across the global media as a violent and atavistic place is the existence of the territorial trap between unionist/loyalist and nationalist/republican places.

The trap, as illustrated in Chapters 1 and 4, is well defined by the walls and other instruments that divide the principal communities within Belfast. The disparate ideological and discursive boundaries between these communities are maintained by a determined lack of interaction across the 'interfaces' that physically replicate these discursive edges. Within segregated spaces many inhabitants aim to maximise their group's status relative to those to whom they are opposed. Given the nature of conflict between these communities it is important to note that the construction and understanding of place-bounded identities are centred upon an introspective contextualisation of 'values' and the historicisation of memory and violent action. As many communities in western societies forget their history or adopt a fatalistic attitude to community decline, elements within the segregated communities of Belfast aim to push memory to the fore in the renewal and defence of their identity. Belfast remains as a place within which the principal identity-formers strive against the arrival of post-nationalist interpretations.

In examining how territorial boundaries work, this book not only explores the meaning of place but analyses the motion across ethno-sectarian boundaries. In many instances this is a study of motionlessness, given that the crossing of ethno-sectarian boundaries is a rare pursuit among the residents of highly segregated places. Understanding the nature of territorial demarcation is important, as it remains the practice through which conflict is performed and indeed assembled. Where political progress has been achieved it has become obvious that it is threatened by the enduring reality of physical separation between sections of the two principal communities.

In order to resolve existing tensions we require, as indicated in Chapter 1, greater deliberation on how territorial boundaries connect with other social phenomena such as demographic decline and access to services. In exploring the nature of boundary maintenance we should also be seeking alternative ways in which to create neutral spaces in order to reduce the physical and cognitive burdens of fear and prejudice. An obvious requirement of change is the capacity to encourage a political imagination that stretches beyond the confines of ethno-sectarian identification. In terms of building democracy, an obvious solution to intercommunity prejudice and violence is to encourage alternative readings of politics within which separation and an ethno-sectarian logic are viewed as repressive and retrograde relationships. The need remains to convince the electorate and younger generations that divisions are not a manifestation of a natural order but are a result of complex, confusing and socially constructed histories. Democracy in Northern Ireland will have arrived when citizens no longer feel the requirement to defend the places within which they live. It would be churlish to suggest that the deconstruction of group- and place-centred identities will be effortless given that territory is a complex, ambiguous and emotional reality of spatial demarcation, but it is a process that is a minimum requirement for consequential change.

The existence of boundaries between communities in Belfast is complicated by various shifts in Northern Irish society. Slow demographic decline within the Protestant community, loss of 'status' within Protestant working-class communities and political growth within Irish republicanism have produced an increasingly negative understanding of 'loss' without deliverance. The shift from a comparatively invulnerable unionism and its related cultural practices to a position within which atavism and insecurity abound has meant that intercommunity violence has appeared as a consequence of that 'despondency'. Whether or not such political and cultural dejection is valid will become one of the key questions that will burden conflict transformation for generations to come. For republicans and nationalists the sense of social marginalisation, unionist cultural 'hegemony' and the 'negative' attitude of unionists regarding their social insertion testify to the failure of unionist leaders to reform the ethno-sectarian disposition of their electorate.

These oppositional discourses regarding what the problems are within Northern Ireland affords those who aim to promote essentialist notions of Britishness and Irishness the opportunity to reproduce ethnic chauvinism and interface violence. We aim to understand the position, fear and susceptibility of both communities with regard to their conception of 'alienation' by advancing important research challenges by means of a diverse yet interlinked empirical design. Chapter 1 explores the meaning of place and the vocabulary of spatial division. In so doing it explores how the reproduction of spatial relationships echoes the existence of social, cultural and economic polarisation. Segregation as a medium is only worthy of study when it is appreciated that the construction of spatial separation is based upon the reconstruction of that relationship. Ultimately, division is not static but is being maintained within a world that constantly alters. Similarly, finding new ways in which to offend and be offended is crucial for the survival of disconnection between oppositional groups.

In Chapter 2 we evaluate the use of an institutional approach, which has dominated attempts to remove conflict and provide for political responsibility within Northern Ireland. There is no doubt that in the period since the paramilitary ceasefires of 1994 the principal aim of various initiatives has been to establish an agreed power-sharing structure. However, institutional approaches are somewhat limited in that they are directed by the conclusion that conflict has been fuelled by inequality and political exclusion, and that the creation of new and inclusive governance structures will alter the nature and meaning of political atavism. Nevertheless, as argued within this book, new modes of governance have redefined the nature of exclusion and have informed the most vociferous interpretations of community exclusion and oppression. The Belfast Agreement's main achievement was to bolster the nature and design of resource competition.

Chapter 2 also indicates that despite a dilution in the meaning of sovereignty there has been a negative counter-position developing around the significance of identity formations that are reproduced through localised meanings and political discord. This has been evidenced by events such as the disputes at Drumcree and Holy Cross,[1] and the perpetual performance of interface violence. As the larger questions of constitutional design and political realpolitik are debated and analysed at the high table of a new political dispensation, the reality of interfacing and segregation has leapt to the fore. It would appear that the everyday consequences of interface violence and restricted mobility were ignored under the assumption that leaders would find a way through the morass of Anglo-Irish history. The realities of several decades of brutal violence, loss and victimhood could not be swept away by Tony Blair's notion that the 'hand of history' was upon us all. Essentially, there were social, emotional and communal costs that were going to have to be encountered before a meaningful settlement could be developed.

Chapter 3 highlights how residential segregation has been a prominent feature of urban division within Belfast since the onset of the industrial age. However, unlike previous periods of intercommunal violence, Belfast has been scarred by contemporary unrest in a different way. A significant effect of contemporary violence has been the virtual disappearance of neutral zones between segregated places. The subsequent increase in segregation, which has been commonplace within parts of Belfast since the late 1960s, has encouraged a succession of bordering events that have extended the connotation and magnitude of preceding ethno-sectarian divisions. The expansion and contraction of ethno-sectarian boundaries have given interfacing a perpetual meaning and form. These borders between unionist/loyalist and republican/nationalist spaces are not simply margins between communities but are crucial mechanisms in the designation of discursively marked space. Interface walls are not merely understood as defensive barriers that impede violence. They are also crucial structures in the lessening of contact between lifestyles and cultural designations that are identified as 'improper'.

The immediate impact of interface walls is to create social, political and cultural distance between communities. The capacity of such boundaries to engender the symbolic signage of cultural and political delineation is both considerable and indubitable.

Interfaces are also an endurable 'aide-memoire' of harm done and of threat unstated. Their existence condenses the performance of violence into distinct space and their locations present a libretto from which community fidelity can be interpreted. Interfaces both annul and confirm intercommunity relationships and constrict space into sites that become the obvious places of violent engagement and intercommunity discord.

Chapter 4 moves the analysis of segregation forward by examining the meaning of physical separation in relation to mobility/immobility between communities. This study is set against the background of an ongoing series of violent acts based upon the intimidation of persons living along the boundaries of segregated communities or those who form the minority within such places. It is abundantly clear that in many arenas vulnerable persons have been beleaguered by physical and emotional violence. However, the most evident impact of violence is to stimulate fears and prejudices in such a way as to further the sense of trepidation, especially with regard to crossing ethno-sectarian boundaries.

The memory and reproduction of violence have had the effect of stopping the desire to cross ethno-sectarian boundaries. Interface violence in whatever form promotes, among those who seek a unidimensional sense of identity, connections between notions of innocence and transgression and debates centred on the concept of 'otherness' within which the 'other' is cast as deviant and malign. Contemporary violence, even though it is on a diminished scale in comparison with violent enactment in the 1970s, still reasserts the desire for boundary maintenance. A meaningful intercommunity dimension will not be established within a context in which interface violence is present, given that each community desires and requests security. Violence has the destabilising effect of undermining the development of normalised intercommunity engagement and mobility between communities.

The fundamental problem is not only that such violence is reproduced. What is more important is that Northern Irish society has failed to appreciate the burdens of segregation upon everyday life. The wider societal implications of interfacing, victimhood and harm have generally been ignored. Undoubtedly, fear has been swept away by a tendency to study politics through an interpretation of political parties and more generalised readings of community. The gritty task of going into communities and discovering what the costs and impediments are has rarely been undertaken.

It is crucial that fears and avoidance strategies between interface communities that have been generated by violence or the acuity of threat are earnestly examined. These societal consequences are key ingredients in the constitution, manner and presentation of ethno-sectarian decisions and loyalties. Avoidance strategies are also of concern because they are real to the people who hold them and they also undermine the ability to create meaningful political stability. Many people in segregated areas are living under a constant sense of besiegement, and thus it is not surprising that they feel that there is still a great deal to be done before the 'hand of history' is lifted from their shoulders. In more general terms, conflict transformation will not suffice if those who suffered most are ignored. As argued in Chapters 6 and 7, policy-makers must plan neutral space and access to such space, in order to break down the burden of sectarianised immobility and the repetitive geography of territorial disputation.

The middle classes and their sharing of space are considered in Chapter 5. There are signs that space is shared in middle-class areas but not in a manner that is convincing. Clearly, there is less hostility in middle-class than in working-class arenas but there are still senses of Britishness and Irishness within the middle classes. More importantly, among those who espouse pluralism there is no overt desire to create a politically identifiable third tradition. As shown, the factor that unites the middle classes most is their belief that house prices will rise. The study of the middle-class group in south-east Belfast produces evidence on the rise of the Catholic middle class. However, a benign sectarianism is also observed in that the Catholic middle classes are younger than their Protestant counterparts and tend to work more in the public sector. Their dependency upon British state-sponsored jobs also seems to temper the desire for a united Ireland. The fundamental point about residential mixing, as studied here, is that it is not driven by a desire to mix but by market forces and housing trends. The middle classes have effectively maintained either a British or an Irish identity and have opted out from supplying alternative political and cultural messages.

There is no obvious post-nationalist turn in Northern Irish society. However, the middle classes provide an example of how British policy-makers were capable of developing some rapprochement between them and the Catholic middle classes. Indeed, the hoped-for outcome was of a Northern Ireland run by the Ulster Unionist Party and the Social Democratic and Labour Party. It is akin to backing the front-runners while taking your eye off those galloping up behind. It is the voters of the highly segregated communities who are crucially important and who must be convinced that they should invest in the development of a new dispensation. In that sense the working class are somewhat different from their counterparts within the European Union in that ethno-sectarianism mobilises them in significant numbers.

There is evidence that the state, politics and policy have become more conscious about segregation but its effects are deduced and interpreted in a highly discerning and at times far from innovative way. While policy-makers are not withdrawing from crucial issues, their ability to launch a coherent and effectual response to the legacy and perpetuation of conflict and the way it is linked with wider social processes has been subdued. The policy landscape is characterised by dispersed and isolated programmes and initiatives, small-scale projects and funding streams aimed at a range of diverse and complex problems. It is evident that each new crisis is greeted by a new programme, often complemented by a new layer of governance and an impenetrable and opaque maze of people, measures and rules to understand and engage with.

Chapter 7 plots micro-programmes dealing with specific interfaces, weak community capacity and environmental management, each with valuable effects but whose overall contribution is blighted by a lack of imagination, legitimacy or vision. These are set against legislative, institutional and resource investment in property capital and attempts to 'shape' Belfast's spatial economy and new consumerist landscape in such a way that it shuns the obvious realism of social exclusion and sectarianism. The problems of segregation are supposedly reconciled via the 'colourless' administrative world of regional and local government, which, in turn, promotes a bewildering series of responses that include denial, problem evasion and the passing of contentious issues to the next most 'convenient' organisation. There are some innovative projects that are linked to local consciousness-raising agendas, but these are exceptional and ultimately untested.

The crucial conclusion from the study of Belfast, as a divided city, is that the issue of ethno-sectarian choice and desire is not 'going to go away, you know'. This is not merely a society condemned by a complex history of irresolvable differences, but a place within which the imagination needed to move forward lacks a platform on which to develop. The idea of turning Belfast into a mini metropole with street-side cafés, postmodern bars and sailing on the Lagan is suspect not only because it suggests a failure to care about those for whom ethno-sectarianism is a burden but also because it highlights a lack of desire to understand what conflict transformation may be.

Setting the city centre up as a swish environment within which to consume and relax will not cover the cracks of a divided city but rather highlight the limitations of political discourse and leadership. A meaningful appreciation of what segregation means and how we can dilute the impact of obvious spatial burdens is the starting point for processes of transition and demographic shift. Maybe we should be reminded of Oscar Wilde's assertion that 'we are all in the gutter, but some of us are looking at the stars'.

1
Even in Death Do Us Stay Apart

Within Belfast City Cemetery there is an underground wall that purposefully separates the Catholic and Protestant dead. In recent times the disputes that have taken place at Carnmoney Cemetery, on the outskirts of Belfast, over demarcating the graves of Protestants and Catholics seem to confirm that even in death there is a desire to remain uncontaminated by the presence of the ethno-sectarian 'other'.

The perpetuation of ethno-sectarian conflict, within Northern Ireland and elsewhere, reminds us that despite the onset of globalisation, cultural homogenisation and mass consumption the links between ethno-sectarianism and residential separation remain central to the logic and explanation of violent enactment and cultural polarisation. The potential of localised, nationalist and anti-pluralist doctrines to determine the reproduction of residential segregation, via particularistic discourses that are constructed around history, politics and culture, remains ever-present (Ackleson, 2000; Agnew, 1993; Bauman, 2004; Delaney, 2005; Newman, 1999).

The problem, when it comes to developing processes of conflict transformation and the development of a shared city, is that local knowledge and experience shape the nature of identity and dilute the ability to present practices that would be capable of shifting the relationship between segregation and political belief (Ashley, 1987; Brenner, 1999; Cohen, 1985). Segregation creates territorial disputes that encourage the development, sustenance and capacity of intercommunal separation and alternative political commitment (Barth, 1969; Ley, 1994; Paasi, 2000). It also presents a medium through which the logic of creating a more settled city is somewhat undermined by the existence and practices of the ethno-sectarian 'other' and/or the state. Identity issues are thus embodied within the definition of place (Lefebvre, 1991; Sibley, 1995; P. Taylor, 1994; R. Taylor, 1988). As Borja and Castells write: 'The creation and development in our societies of systems of meaning increasingly arise around identities, expressed in fundamental terms. Identities that are national, territorial, regional, ethnic, religious, sex-based, and finally personal identities – the self as the irreducible identity' (1997: 13).

Theorists of various forms of segregation are increasingly, and correctly, concerned with the interpretative nature of urbanism and discursive practices. Harvey (2000), Imrie (2004) and Marcusse (1993) in particular have highlighted the need to ground 'lived experiences' in order to understand the subjective interconnections and emotional forces that drive the urban 'moralities' of segregation. Thus, the need remains, for those who study segregation, to locate the structural relations of power, displacement and conflict, which determine how spaces are both (re)constructed and contested (Wasserstein, 2001). This ultimately means determining the mediums through which Belfast's residents transform daily occurrences and emotions into a symbolic system of territorial attachment (Giddens, 1991; Gottman, 1973; Ley, 1994). These constantly negotiated and contested social and spatial practices matter in that they are interpreted and given significance by their participants (Shirlow, 2003a, 2003b), in what Jackson (1989) refers to as the capacity of social groups to develop distinct patterns of life.

Within the Belfast context the desire to locate the 'self' within a social, cultural and political group, via a combination of imaginings and experience, is attached to what Sibley (1995) identifies as a process of 'cultural production' and the formation of boundaries between the 'chosen' and the 'rejected'. Identity construction is generally based upon relational concepts as opposed to objectivity, introspection and rationalism (Mandaville, 1999; Sack, 2003). In terms of space, the formation of social, cultural and political beliefs crosses spatial scales including residential segregation, schooling, leisure, sport, consumption, socialising and the workplace. As the information presented within this book indicates, a central goal in the exploration of segregation is to determine not only the nature of contact between spatially separate populations but to also designate how ideas, beliefs and behaviours are reinforced by their social milieu (Dumper, 1996; Killias and Clerici, 2000).

There is no doubt that the casting of Belfast as a 'normalising' place sits in stark contrast to the realities of ethno-sectarian separation, multiple forms of segregation and the reproduction of violent acts and cultural opposition. The capability of localised and sectarian doctrines to reproduce residential segregation remains linked to the presentation of competing ideologies that are influenced by discourses of 'truth', mistrust and intercommunity misunderstanding (Maguire and Shirlow, 2004; Shirlow, 2003a; Shirlow and Murtagh, 2004).

SPATIAL BORDERING

The ever-present conflict within Northern Ireland is not based upon religion but rather religion acts as a boundary marker with regard to competing aspirations regarding forms of Britishness and Irishness. As Jenkins concludes: 'Although religion has a place in the repertoire of conflict in Northern Ireland, it is apparent that, for the majority of participants, the situation is seen to be primarily concerned with matters of politics and nationalism, not religious issues' (1986: 16). The border between Northern Ireland and the Republic of Ireland remains as the key, symbolically defined feature of cultural and political discord (Anderson and O'Dowd, 1999; Wilson and Donnan, 1998). The constitutional border is replicated within cities, towns and villages throughout Northern Ireland. Conflict and its reproduction, at the micro level, are linked to the symbolisation of a constitutionally divided island.

In Belfast, unlike most divided cities in the European Union, the most acute and perceived spatial divisions are not simply those of class or race but those of national identity. For republicans and Irish nationalists, the reproduction of segregation is tied to a series of acts and discourses within which the British state and hegemonic unionism organised space in order to control ideologies of resistance and the praxis of dissidence by 'disloyal' citizens. The ability to express Irishness through the removal of the prohibition on hosting the Irish national flag in the wake of the Anglo-Irish Agreement (1985) and the pursuit of Irishness through language, dance and theatre, within republican/nationalist[1] places, have been linked to a desire to alter the symbolic nature of space.

Identity formation has not been free-flowing but tied to distinct practices of seeking a form of Irishness that has been denied, rejected or forgotten. The repossession of Irish symbols is not a benign process of reclaiming but a defiant act of community-based veneration and a challenge to other symbols and political codes. For loyalists and unionists the state is seen as insufficient and culpable in the rise of Irish republicanism, and as a result the maintenance of 'Protestant' places is tied to a singular preservation of nationhood and the 'British' way of life. Demographic decline within the Protestant community has also created an enduring sense of decline and ideological and cultural defeat. These two main ethno-sectarian blocs have over the past 30 years drifted apart owing to violence and ideological intent. As a result of the separation – a process that Gellner (1986) understood as being centred upon differences becoming so crucial that they became reflected in people's emotional attitudes – we are left with a city burdened by atavism and institutionalised identities.

The ethno-sectarian problems that exist are not simply based upon distinguishing forms of national identification and their organisation within space. Nor is it the case that spatial distinctiveness exists precisely because political beliefs do not complement each other. A central reason why spatial disunity exists instead of a more pluralist and reasoned sharing of space is that the nationalist/republican community is engaged in a form of political ascendancy while the unionist/loyalist community struggles to maintain unity, purpose and spatial well-being.

Each community that resides along an interface between oppositional spaces lives with the fear of attack. However, many predominantly unionist/loyalist places are burdened by demographic decline. This latter reality creates unambiguous fears concerning wider sectarianised notions of territorial dissolution within the unionist/ loyalist community. The twin processes of renewal and decline that define division in Belfast are part of a complex set of issues that reasserts the power and authority of segregation. In sum, segregation exists because new reasons for the existence of such separation, such as the fear of territorial loss and the contestation over parading, are found and acted upon. Segregation, in Belfast, is not based upon an ethno-sectarian standstill or merely the maintenance of frontiers but upon the need to re-deliver the meaning of separated living through novel, as well as rehearsed, narratives of inclusion, practice and belief.

The issue that affects both communities is not merely that the border exists but that the survival of the border is challenged. The shift from a relatively secure unionism and its associated cultural practices to a position within which republicans and Irish nationalists continually gain demographic, cultural and political power has encouraged a sense of fatalism that is reproduced via atavistic attitudes within sections of the unionist community. For republicans and nationalists, their sense of social marginalisation, unionist 'triumphalism' and the 'objectionable' attitude that unionists have regarding their civic inclusion testify to an irreformable disposition within that community. In sum, and despite the 'peace process' and the sometime delivery of devolution, the desire to hunker down behind essentialist notions of Britishness and Irishness and related forms of ethnic chauvinism remains ever-present.

Segregation is one of many manifestations of disagreement and a failure to create a version of Northern Ireland that is acceptable to the inhabitants of that place. The continual reproduction of violence between segregated places and the recent building of new interfaces testify to the strength of segregation in constantly shaping politics within Northern Ireland. The significance of segregation was clearly stated by Darby:

Just as one cannot hope to understand the Northern Ireland conflict without an acquaintance with its history, it is impossible to appreciate its pervasiveness without some knowledge of the background, extent and effect of residential segregation between Catholics and Protestants. This is both the cause and consequence of the province's history of turbulence. (1976: 25)

Segregation has had the effect, via spatial bordering, of creating containers within which many residents have, in the main, come to accept a unidimensional cultural and political definition. These are also places from within which communities can articulate and emphasise the issues of discrimination, inequality and the denial of their human rights. Furthermore, segregation has been linked to the presentation and enactment of violence. However, a central issue is that segregation and the related control of territory create approval and justification for separation. Segregation provides the ability to invest faith within place and to acquire cultural companionship. In effect, it creates a sense of homeliness within which residents imagine themselves to be at one with their neighbours and community. Discourses of shared values, self-worth and identity within segregated spaces are crucial determinants in the perpetual dilution of a shared common ground within Northern Irish society.

In crude terms segregation exists because it works. The perpetuation of segregation is based upon a process of operationalising the difference between republican/nationalist and unionist/loyalist spaces. This is achieved through casting segregated places as more homogeneous in social and political terms than they actually are. Loyalty to place unites the atheist with the godly, the leftist with the right-winger and the male with the female. It has the capacity to disguise differences through achieving a sense of value and attachment to a struggle over the maintenance of places that must remain unique. Belfast remains not as a city but as an assemblage of 'villages' within which detachment from other places is crucial in terms of identity formation. The political divisions that constitute conflict within Northern Ireland could not achieve such a form of representation without segregation and the controlling effect that separation has upon 'lived' experience.

There is no defined or long-term political desire to deconstruct the reality, nature and reproduction of segregated living beyond the ultimately depressing belief that benign forms of segregation will suffice. It is evident that the presence of segregation, even within a less violent disposition, militates against the delivery of long-term political, social or cultural stability. This has been proven throughout the last decade of 'peace-building' by the events that surrounded the Holy Cross dispute, Drumcree and the political fall-out following the Northern Bank robbery. The reproduction of such political atavism is linked to the capacity of segregation to deliver meaning to what are substantial but alternative political mandates.

Segregated communities are places that have been engineered, fabricated and managed by political entrepreneurs seeking to mobilise political discourse through territorial control. Such places can be described as 'sites of resistance' that have emerged where cultural, economic and political differences have interlocked in order to produce fractal spaces that are characterised by concentrations of loyalty to place. As Heikkila argues, this conception of space represents the key challenge to those wishing to deconstruct the power of cultural opposition: 'Space matters because it mediates the experiences of people in places, and further, it shapes the structure of the opportunity set available to them' (2001: 266). In analysing the impact and effect of segregation and of bordering between communities, it is asserted that the type of radical political adjustment needed to positively alter the nature of ethno-sectarian practice remains underdeveloped. It is argued that given the nature of residential segregation and the impact that it has upon spatial mobility the capacity for 'normalised' social relations remains, in the short- to medium-term, unlikely.

SPACE (REALLY MATTERS)

All phenomena are temporal, and without doubt conflict within Northern Ireland is infused with history, but the playing out of conflict has also happened in space and therefore conflict also has a series of geographical dimensions. In terms of academic deliberation there has been insignificant attention given to geographical analysis and the exploration of conflict has undervalued the importance of obvious spatial relationships. Segregation is one of Northern Ireland's strongest spatial determinants, representing the limit of political, cultural and social confinement, and the capacity to displace the development of alternative notions of political belonging and belief. Segregation is not merely an inherited structure that undermines a new political dispensation. Rather it is a form of spatial practice that is constantly reproduced as a 'proven' system that encourages struggles over urban and rural spaces. Such spaces are both physical (built environment) and symbolic (prejudice and fear), as well as a combination of personal space and community experience.

Boal (1969, 1976, 2000) conceptualised responses to conditions of ethnic–religious segregation and cultural decline in terms of a continuum of 'loyalty, voice and exit'. Members of communities who feel isolated or threatened may remain loyal to the area in which they live and try to exist in as secure a position as possible. They may voice their concerns in a number of ways, such as campaigning or demonstrating or even through violence. Others choose to exit and simply leave the locality and seek sanctuary within their own ethno-sectarian group. Historically within Belfast, both responses can be identified. The latter trend has generally been understood to be due, in the main, to the outmigration of Protestants from the city. The catalyst for such outmigration was usually de-industrialisation, increased violence and a subsequent de-territorialisation of spatially vulnerable populations. Conversely, the maintenance of small Protestant and Catholic enclaves, especially Suffolk and Short Strand, points to a form of spatial loyalty and perseverance.

There is a growing stability in terms of aggregate population trends within both communities in the Belfast Urban Area. The opening up of new spaces for housing development as a result of market forces, demilitarisation and the recent history of Protestant population decline sustains a sense of spatial anxiety among many. For republicans, those most likely to be on housing waiting lists, the failure to accommodate their communities' housing need is viewed as being based upon state policies that purposefully sustain unionist/loyalist communities that are demographically unviable.

Residential segregation is linked to wider processes of territorial marking, which include wall murals, flags and kerb paintings. While the contestation over space and territorial control is not necessarily linked to population decline and territorial deterioration it is evident that the commemoration of identity aims to provide some certainty in the context of political change. In addition, residents of segregated places adjust their travel patterns and use of services and facilities in response to fear and perceptions of threat from the majority (Shirlow *et al.*, 2003). An overall problem is one of how both communities 'voice' legitimate concerns in a manner that is not read as being aggressive or linked to crude prejudices.

'Exit' from insecure circumstances has often been the principal response in both Catholic and Protestant areas of Northern Ireland. For example, exit of Catholic populations has been characteristic in parts of the Moyle, Ballymena, Carrickfergus and Larne District Council areas in recent years. Poole and Doherty (1996) argue that Protestants, in particular, have made slow micro-adjustments to ensure the maintenance of their majority numbers in particular places. Using intercensal data, they demonstrate that this has particularly affected the western region of Northern Ireland and border areas. However, what is particularly damaging about this pattern is its selective nature. Murtagh (2002) shows that it has tended to be the younger, the more mobile and the employed and employable who have left first. Exit can therefore begin a process of residualisation whereby the remaining minority is likely to be comprised of older, benefit-dependent and less socially or spatially mobile people, which further complicates the task of community development and renewal. Such 'hollowing out' can dilute the capacity to develop social capital and the skills needed to create new participatory modes of engagement with the 'other' community (Murtagh, 2003a, 2003b).

The existence of vulnerable places is configured by a series of cultural, political and social relationships. Ultimately, divided space produces alternative forms of social interaction and sustains the politics of opposition. Spatial fetishism furthers the capacity to manipulate space and create micro social orders defined by republican/nationalist and unionist/loyalist logic. Space is not merely a container of separated political ideologies but also a producer and cause of social relations. Segregation not only engenders political separation but also enhances societal distinctions between physically separated peoples.

Within the conflictual arena that is Belfast many citizens have sought to alleviate their fears and prejudices and in so doing mitigate the incidences of harm upon them through defining a sense in which segregated living is protective given the physical reality of walls and interfaces between them and the 'other' community. Lifestyle is also, usually through restricted spatial mobility, altered in order to stop entry into places dominated by what is viewed as a community of harmfulness. Such architecture of fear and prejudice is complemented by living within segregated communities that articulate an internal perspective regarding such cognitive 'reasoning'. Segregated places aid the fashioning of community belief and the certitude of mutual solidarity and spatial control. Interfacing encourages enclosure of ideas, the monitoring of the entry of strangers and the symbolic exclusion of the politically undesirable.

In effect, segregation provides the base and rationale upon which paramilitarism can flourish given that paramilitary groups act as communal defence forces. The enclaving of social and cultural activities and the claiming of space create a process whereby those who are oppositional become external, distant and remote. However, enclaving and separation do not fully protect as the creation of segregated space requires modes of defence. Ultimately the practice of violence remains as segregation creates deep forms of spatial isolation and mistrust. Interface walls between segregated areas are not merely protective barriers but visible representations of extreme and divergent political codes that are inscribed upon place (Delaney, 2005).

The differences between respective populations, that segregation illustrates, are reinforced by inequalities between each group, both real and perceived. The relatively high rate of Catholic unemployment and the increasing incapacity of many Protestant places to maintain demographically even populations create a further series of conflicts concerning the state's role in fairly maintaining the lifestyle and social well-being of both communities. Thus unemployment and social exclusion are understood not merely through social class but via community-based experiences and structures.

The negotiation of difference with regard to segregated urban space is articulated through the casting of the 'other' community as symbolically impure. The question of who is felt to 'belong and not to belong' (Sibley, 1995) shapes segregated space and in so doing limits the capacity to mix and to create alternative experiences that are not burdened by mistrust and near paranoia. Thus segregated spaces are representational and designed through exaggerated notions of social difference and complex patterns of exclusion that are tied to those who conform to defined symbolic and cultural codes within the 'home' territory.

POLICY LACUNAE

The use of public funds to build defensive walls and maintain buffer zones between segregated places deepens the meaning of segregation by reducing mobility and contact between diverse populations. Such strategies also imply that the state has failed to provide public spaces that are safe for all. The state has accepted segregated ethno-sectarian design over the creation of democratic values that could be played out within a mixed community-based spatial framework. The inability to protect led to a series of policies of conflict management and the building of 'defensive' walls between communities. The creation of such defensive features underlined the failure of state policy owing to the inability, despite physical separation, to insulate those who lived in such places from harm.

The development of policies of physical separation encouraged forms of spatial exclusion, the defence of place and the promotion of unidimensional cultural values and wider strategies of political resistance. The deepening of segregation through violence and policy remade established segregated space and in so doing deepened socio-cultural divisions.

With the onset of extreme violence in the late 1960s, new versions of dangerous places and moral panic were intertwined with established suspicions regarding the threat posed by the 'other' community. Even though contemporary political violence can be used to justify the perpetuation of segregation it is clear that violence is one of several factors that account for this social and cultural practice. There is no doubt that during the 1970s and 1980s segregation was intensified by the impact of violence (Whyte, 1990; Wilson, 1989). The increase in segregation at that time had the effect of conditioning a series of emotional and cultural concerns between communities. Anxiety has been an obvious outcome given the history and reproduction of violence between segregated areas. Phobia, insecurity and uncertainty remain owing to low levels of social trust and an inability to control the nature and direction of violent acts. Fear and mistrust are linked to understandings of risk, doubt and behavioural responses that are influenced by the assessment of threat. Feeling threatened may not be directly related to levels of violence, given that interpretations of risk and threat can be more extreme than contemporary levels of actual violence would predict. Among some, these prejudiced attitudes towards the 'other' community are influenced by the representation of knowledge and the collectivisation of memories in a desire to present the 'other' through discourses of danger and impurity.

Evidently, space matters when it comes to understanding the perpetuation of violence and political discord. Extreme links between ethno-sectarianism and place have always been tied to the maintenance of residential boundaries and the marking of such boundaries through both symbolism and force. Space thus represents the crucial category for understanding the connections between all sorts of multiple oppressions and the possibility to mobilise around agendas framed by place (Shirlow and Shuttleworth, 1999; Shirlow and McGovern, 1998). The social production of space is as described by Soja: 'an important strategic milieu for a new coalition politics of class, race, gender, sexuality, age, ethnicity, locality, community, environment, region and other sites and sources of cultural identity and the assertion of difference' (2000: 216).

Segregation influences the localised nature of the politics of territorial control and resistance, where the authority of communal separation, segregation and exclusion prevails over the politics of shared interests, amalgamation, incorporation and compromise. The relevance of ideologically constructed space in the reproduction of politicised identities, which are linked to notions of 'besiegement', cultural dissipation and fear, is crucial in interpreting the link between space and social, cultural and political reproduction. Despite significant political morphology, the technical instruments of ethno-sectarian discord are still being reproduced within what are termed 'interface areas'. In those arenas, predominantly socially and economically deprived, in which unionism/loyalism and republicanism/nationalism are separated by both physical and mental constructions violence and mistrust between communities are ongoing and prejudice towards and avoidance of the 'other' community, through a combination of fear and prejudice, are commonplace and daily occurrences.

Fear and the link between it and residential segregation impede the search for work, the uptake of training and education and the use of public services. In addition, fear creates socio-spatial burdens that are endured by socially deprived communities. The potential to reconstruct Northern Ireland's production and consumer arenas, in order that they respond to equality of opportunity and parity of esteem, is a major factor in the creation of long-term political stability. Few if any of the policies that aim to challenge socio-economic dislocation are linked to the realities of spatialised fear and the reproduction of social deprivation and communal polarisation.

Even though the Belfast Agreement identified the spatialisation of fear as a policy priority for a number of territorial and economic development-based departments, there is no evidence that it has become a key issue in policy formation. This is not simply a matter of data and methodological gaps but because of political 'sensitivities'. In reality most public sector agencies have actively played down the interplay between rigid residential segregation, rising sectarian tensions and ethnic territoriality. Downplaying the nature of social and political exclusion remains a feature of neo-liberal governance. Furthermore, the lack of consensus among political parties and the nature of sectarian disputation that exists between them suggests that the ability to articulate intercommunity concern is severely hampered.

Where strategic policies have emerged, they have tended to be attached to the central features of crisis management. The key principle of crisis management has been to imagine that abnormal problems can be solved via normal policy-making strategies. Evidently, the desire to collect information on the degree and nature of sectarianised habituation remains negligible, especially given that the reading of such data would be undermined by subjective interpretations. The lack of significant policies directly linked to recognising the existence of sectarianised fears suggests a form of 'policy blindness' and a refusal of government departments to challenge the edifice of sectarian opposition.

The analysis of fear and prejudice, as provided within this text, underpins the need to translate into practice policy and political rhetoric around segregation and to connect practice with the reality of sectarianised habituation. Given the sectarian nature of political control in Northern Ireland, much of the normative commentary dedicated to challenging segregation and ultimately the reproduction of fear and prejudice is ambiguous and difficult to distinguish in the detail of policy-making instruments.

DIFFERENCE BETWEEN; DIFFERENCE WITHIN

As argued by Horowitz (1986) and Smith (1991), much work on ethnic segregation tends to overplay the cataloguing of peoples and ignores the nature of political heterogeneity within ethno-sectarian groups. It is important to note that the violent, cultural and political acts that aid the reproduction of segregation should not be read as being supported by all residents of segregated communities within which intercommunal violence is of concern. Such an interpretation does not deny the impact of segregation upon political belief but signifies that attitudes within segregated places can vary. Moreover, the reluctance to enter areas dominated by the 'other' ethno-sectarian group can be influenced by threats, both imagined and real, that are made against people by members of their 'own' community. Segregation should not be understood as simply a set of atavistic relationships 'between' places but also to exist 'within' places themselves.

The construction of ethno-sectarian landscapes is influential but does not make everyone who lives in such places accept homogeneous senses of belonging and affiliation. In simple terms, many people who live in highly segregated communities do not support a unidimensional notion that their dreads and prejudices are sourced merely from elements that exist outside their 'own' community. This does not mean that such persons are completely against the ascendant political representations that exist within their own community. Instead it could be argued that highly segregated communities contain diverse populations that reject, partly reject/ accept or accept symbolic representations and discursive hegemonies that are related to wider ethno-sectarianised discourses. It could also be argued that devotion to and rejection of ethno-sectarian discourses are temporal and influenced by the appearance of violent episodes, the marching season and a series of political vacuums caused by the ongoing collapse of the Northern Ireland Assembly.

The existence of those who do not ascribe to sectarianised modes of habituation and affiliation indicates that ethno-sectarian boundaries are less fixed than has been assumed. However, somewhat crucial to the understanding of conflict transformation in Northern Ireland is our argument that the reasons for ethno-sectarian separation are not merely framed by the logic of intercommunity relationships but also by the belief, held by non-sectarians, that the pursuit of intercommunity activities is unwise owing to intracommunity-based threats. The construction of sectarianised affiliation casts those who do not accept cultural affiliation and repetitive ethno-sectarian discourses as both deviant and treacherous. As the former MP Gerry Fitt used to say, 'If you stand in the middle of the road you will get knocked down.' Our work indicates that the implied de-territorialisation needed to shift Northern Irish society towards more agreed and agreeable forms of political ownership and consensus-building remains distant and at present geographically rootless.

Spatial examples of integration, mixing or even coexistence are sparse, but they are there and distinctive urban forms have merged in the economic transition that has accompanied the peace process and post-conflict opportunity. Macro-indicators of economic performance show remarkable growth in gross domestic product (GDP), employment and house prices (Murray and Murtagh, 2004). But, at the same time, welfare benefit dependency and income poverty have increased for a smaller number of the population. This emerging 'dual economy' is increasingly spatialised as new sites of consumption in the city centre, river front and former docks have replaced the traditional productive economy centred on heavy engineering and shipbuilding. Those with skills, education and access to finance have done well in the new economy, while those without resources are increasingly corralled in the 'sink' estates of the inner and outer city.

The correlation between segregation, fatalism and economic disadvantage has opened up a new socio-spatial fracture in the city, which is becoming increasingly important in shaping people's life chances, educational competence and employment opportunities. The 'other city' – the new suburban developments, with their gated qualities – reflects this post-industrial reality and can be found and read in similar ways to any other late capitalist urban system. But in the context of Belfast, its emergence is empirically interesting and raises important research questions central to this book. Does it really represent a new spatially differentiated class with distinctive interests shaped by material pursuits and not simply atavistic sectarian claims? How integrated are they and are these places where ethno-religious identity matters less or not at all? Can they provide lessons for the transfer of experience or practice in the creation of shared housing options, which the various Life and Times surveys inform us, are the preferred neighbourhood for more than 60 per cent of the Northern Ireland population?

In this book we look in particular at how the city functions, how spatial relationships and politics are shaped by its ingrained territoriality and at the internal cohesion and fracture in localised communities. Ideas around social capital have become important in determining the structuration of spatial communities in both working-class and middle-class spaces. We use empirical analysis to look at the distinctive patterning of social activities and relations in suburban South Belfast in order to define distinctive processes around the emergence of a new consumption class. A complex picture emerges in which middle-class pursuits certainly moderate attitudes, values and behaviour but fundamental ethno-sectarian subcultures dominate socio-spatial relations. It is important not to overstate the extent of integration in mixed housing spaces but it is also important not to underestimate the capacity for people, with multiple identities, to share territory and resources in Northern Ireland.

WHOM TO TRUST?

The capacity to desegregate residential districts within Northern Ireland requires full political cooperation across the spectrum of political parties, agencies and the two states. Desegregation must emerge as a key component in the rebuilding of community lives that have been shattered by both contemporary violence and the history of modes of segregation that were tied to population influx in the nineteenth century. There is no doubt that individuals and groups, many of whom were involved in paramilitarianism, are now working, via community organisations, to build some form of intercommunity linkage with the aim of reducing the potential for future violent acts.

Yet such efforts do not challenge wider narratives of resistance and interpretations of segregated living that point to the legitimacy of segregation as a rational mode of habituation. Apologies, an important part of conflict transformation, have emanated from within loyalism, republicanism and the British state but such 'repentance' is usually tied to a selective understanding of victims or victimhood. The growth in demands for inquiries and truth commissions highlights the reality that victimhood remains a symbolically charged ideological battleground. Politics, despite change, still works upon the mobilisation of unionist/loyalist and republican/nationalist communities through exploiting the narratives of violence and harm. Within such an intensely divided society it is evident that mobilising harm, via commemoration, wall murals and gardens of remembrance, is undertaken through recognition of the hurt endured and more importantly acknowledged within selective community memories (Graham and Shirlow, 2002).

The media and the British and Irish states have aimed to censure the violent behaviours that have emerged from Northern Ireland's most politicised and violent communities. The violence that continues to happen within republican and loyalist communities continues to be denounced as deviant, pathological and criminal. For Irish republican groups, their engagement in violence was viewed as 'necessary' in the protection of republican areas and the development of 'anti-state' strategies. For such groups it was/is the violence produced by the British state and loyalists that is viewed as illogical and based upon illicit depictions. Loyalist paramilitaries argue that their violence was based upon a determination to shield loyalist places and a 'legitimate' and historically constituted right to safeguard their political beliefs.

In addition to this the British state remains generally silent when it is cast as an abuser of force or as a collaborator in violent enactment.

The ultimate issue is to identify those who are to be trusted with the task of delivering an unfettered politics of renewal, trust and intercommunity dialogue. The capacity to shift the electorate on to a common plain of understanding and interpretation remains undermined by the history of such agents in promoting discourses of cultural and political homogeneity. 'Truths' were told and explained, by those with influence, within the context of the binary argument that the 'self' was trusted, normal and victimised whereas the 'other' community was abnormal, loathsome and diffident. To break such an enduring mould of thinking can only be achieved through a self-reflective critique that challenges the central pillars of self-belief and political worthiness.

The casting of blame upon others remains central in the definition of political action and commitment. Loyalty to the community of the 'self' sits high above concern for those who are opposed. This is not simply the politics of futility or atavism but it is the central force that defines political 'logic' and the holding of loyalties of identity. Identity is infused with imagination, but more important than that is the functioning of identity within defined material, social and cultural practices.

The desire to champion political discourses and violence as 'veritable', 'virtuous' and legitimate has led to a conspicuous failure to recognise how ethno-sectarian conflict has reproduced spatial enclosure and behavioural practices that are partly dictated by widespread fears and prejudices. Presenting the 'other' community as fearsome and pathological was employed as a tactic that aimed to homogenise communities and reinforce spatial enclosure. Spatial enclosure via segregation was not merely about creating boundaries between alternative ideologies but also concerned with building relationships within segregated places that promoted resistant cultures through the development of an unwavering community-based political logic.

Significant sections of both communities championed the case that they were the 'bona fide' victims of violent conflict. This refusal or inability to identify that harm was caused to those cast as antagonists remains a prevalent feature of political argument. Accepting the victimhood, especially publicly, of the 'other' community is considered politically inappropriate within the context of ethno-sectarian opposition, as it would emasculate ideological genuineness and encourage a process of accepting responsibility for violent acts. Such a political shift would incapacitate the legitimacy of political 'resistance'.

It is important to stress that the decrease in politically motivated violence over the past decade has been accompanied by an intensified process of claiming and 'owning' victimhood. Despite the fact that such a political logic undermines political progress, claiming the totem of having been the most persecuted community undoubtedly assists in perpetuating the rectitude of political justification. The reality is that political righteousness has been tied to sectarian separation, and residential segregation has meant that there is a need to present those opposed to your political logic as being motivated by a desire to reject meaningful political progress.

In terms of the representations that emerge from the media and the British and Irish states, the fact that many people, often living in the most socially deprived and violent communities, felt unprotected, fearful and threatened has been conveniently ignored in the profiling of communities as either 'deviant' or 'trustworthy'. The presentation of those who were labelled as deviant and transgressive cannot now be counterposed by an acceptance that those cast as 'deviants' were themselves victims of politically motivated violence. The whole edifice of political belonging would be severely tested if communities and the British and Irish states were to accept an alternative belief system, one that recognised that the ideologically constructed 'bad' suffered at the hands of the discursively constructed 'good'. Remaining blameless thus remains a key component in the perpetuation of sectarian atavism in Northern Ireland.

In the context of the peace process and more recent violent events it is clear that the comprehensive desire to circulate an image of stability means that those engaged in violence must remain categorised as an atavistic cabal intent upon destroying political 'progress'. Recognising that many people involved in ethno-sectarian and violent acts experience disproportionately high rates of fear and exposure to risk is rarely denoted as a motivating factor for the perpetuation of intercommunity discord.

STUDYING AT THE INTERFACE

In his seminal work on ethno-sectarian division in Belfast, Boal (1969) provided a range of crucial insights into how political violence and the politicisation of space determined the nature of religious segregation in Belfast. His work was significant in that it examined how long-term historical structures and their enduring memory influenced and directed the shaping of ethno-sectarian enclaving within Belfast's contemporary urban system (Boal, 1976; Burton, 1978; Poole and Doherty, 1996).

An extensive tradition of research into the effects of conflict in Northern Ireland on spatially segregated communities exists. The body of geographical research has concentrated on the statistical descriptions of the levels of segregation, the sociological construction of territorialised places and the effects of segregation on enclave urban and rural communities. Hargie and Dickson (2003) describe the range of research conducted on community relations in Northern Ireland and identify the need for sensitivity, objectivity and rigour in researching the impact of ethno-social division. In particular they emphasise the need for, and importance of, mixing methods to reveal the wider complexities of segregation and to determine how divisions are reproduced in the built environment. Murtagh (2003b) used quantitative survey approaches to map out attitudes and behaviour of people living in interface areas around Belfast while, conversely, Connolly and Healy (2003) contend that quantitative analysis alone cannot unpick the processes at work in divided communities and among specific groups who feel alienated. They argue that only qualitative approaches can fully capture the lived experiences of divided cultures and uncover the causal relationships that explain why identity groups act in the way they do.

This book combines both qualitative and quantitative methods in what Hoggart *et al.* describe as 'triangulation', which means:

the use of a series of complementary methods in order to gain a deeper insight on a research problem. The advantage of using complementary methods is that they enhance capacities for interpreting meaning and behaviour. This is because the insight gained can strengthen confidence in conclusions by providing multiple routes to the same result. (2002: 212)

Similarly, Connolly argues that researchers in Northern Ireland:

should be committed to the unbiased and objective pursuit of knowledge. They have a responsibility to report their research comprehensively and accurately, including the methods they have used and the data they have gathered. Researchers must avoid selectively reporting their findings or fabricating, falsifying or misrepresenting their findings in any other way. (2003: 6)

Connolly has developed criteria for researchers dealing with vulnerable groups, which include the need to:

- conduct their professional work with integrity and in such a way as to not jeopardise future research, the public standing of researchers or the ability of others to publish and promote the findings of their research;
- respect the rights and dignity of all those who are involved in or affected by their research;
- ensure as far as possible the physical, social and psychological well-being of all those who take part in their research or are subsequently affected by it.

This book draws on these broad principles and combines quantitative and qualitative traditions to construct a deeper picture of segregation and its imprint on the city. Small-area statistical analysis and modelling are combined with quantitative household surveys and in-depth interviews to broaden our understanding of the interplay between attitude, experience and behaviour in the maintenance of segregation and the possibilities of mixing. This work indicates that ethno-sectarianised fears and prejudices in relation to the 'other' community play a decisive role in the choice of arenas within which to consume and undertake leisure activities – a classic case of data demonstrating extensive patterns of polarisation between the communities studied. The interviews challenge simplistic assumptions concerning spatial interaction and suggest that previous understandings of residential segregation are somewhat underdeveloped.

2

The Belfast Disagreement

An institutional approach has dominated attempts to build peace and political accountability within Northern Ireland. The central aim of these initiatives has been to re-establish and further the power-sharing structures that aimed to increase cooperation between political parties as outlined in the Sunningdale Agreement of 1973. Institutional approaches are guided by the supposition that conflict has been fuelled by inequality and political exclusion, and that the creation of new and inclusive governance structures will transform the nature and meaning of political hostility. The recent pursuit of new modes of governance has redefined the nature of exclusion and bolstered the political fortunes of those political parties that deliver the most vociferous soundings regarding their community's exclusion.

It is also maintained that division can be healed through the adoption of new power relationships that followed the Maastricht Treaty and the decline in the centrality of sovereignty within the definition of nationhood. Without doubt, the British and Irish states have aimed to dilute their claims of sovereignty regarding Northern Ireland. Such changing circumstances have inaugurated a more definite role for the electorate in terms of defining the constitutional future of Northern Ireland. However, the alteration in constitutional design, as evidenced by devolution within Northern Ireland, has not significantly modified the meaning and significance of identity formations that are reproduced through localised meanings and political discord.

Institutional fixes and the political programmes that have been advanced remain tied to alternative strategies of conflict management and more inclusive and creative processes for change. Devolution has been a central plank within the new political dispensations as it has consolidated a 'new' Northern Ireland within which local politicians have gained increasing control over the mechanisms of governance. In addition, potential unionist dominance has been replaced by power-sharing and an institutionalised, quasi-sovereign role in Northern Ireland for the Irish state through the North–South ministerial council.

Disagreement over the naming of the Agreement, adopted on 10 April 1998, which led to devolution, exemplified the uneven ground upon which a lasting settlement was to be built. Unionists prefer the appellation 'the Belfast Agreement' while nationalists/republicans use the term 'the Good Friday Agreement'. The use of different names for the same object is of course commonplace within a divided community, with other notable examples including the use of Derry or Londonderry and the term Six Counties by nationalists/republicans to describe Northern Ireland. Divisive naming is symptomatic of the perpetual reality that Northern Ireland, despite significant and meaningful political change, remains as a 'disagreed' place within which politics remain disagreeable.

The meaning of the Agreement is interpreted through diverse rationales of constitutional change. For unionists the legitimacy and integrity of Northern Ireland have been underlined by devolution and the endorsement of the principle of consent, which upholds the majority's objective of remaining within the United Kingdom. Conversely, among nationalists and republicans the creation of a power-sharing executive and the creation of cross-border bodies have been identified as the beginning of the end for Northern Ireland as a constitutional entity.

The inability to maintain devolved structures has also been interpreted via the disparate lenses of unionism and nationalism/republicanism. For unionists it was the failure of the PIRA to disband on 'their' terms as well as allegations concerning that organisation's recruitment of volunteers and criminal activity that undermined the capacity to develop and bed down new institutions of governance. The emergence of Sinn Fein (SF) as the dominant voice of northern nationalism has furthered fears regarding the desire of their electorate to move beyond 'duplicity' and 'untrustworthiness'. For republicans and nationalists, it is the unionists who fear change and who block the development of social, cultural and economic equality and parity of esteem.

Between the pillars of acrimony and triumphalism lie the ruins of what could loosely be described as a middle ground. An evident feature of the hostility between the Democratic Unionist Party (DUP) and SF throughout the experiment with devolution has been the erosion of a centrist middle ground overwhelmed by the expansion in electoral shares held by political parties that uphold the most apparent and vociferous notions of cultural identification and political triumph. Somewhat ironically the accommodationist middle ground has been diminished within a series of structures and initiatives that aimed to support the development of such a community. As noted by Wilson:

More generally, the communal registration process has militated against the emergence of a strong political centre that might engender stability in the institutions. Indeed, on the contrary, it has reinforced 'groupist' stereotyping characteristic of media reporting in Northern Ireland, where actual Protestant and Catholic individuals are constantly hoovered up into ethnonationalist 'communities' belying the pluralism of real social life. (2003: 15)

Wilford and Wilson have argued that sectarianism was also entrenched through an either/or constitutional choice between the Union and Irish reunification and the single-transferable vote electoral system for the Northern Ireland Assembly (NIA). For them the fundamental problem within consociationalism is that it is premised upon the divisions it aims to resolve. It

...assumes that identities are primordial and exclusive rather than malleable and relational: high fences in other words make good neighbours! A fundamental condition of consociationalism is an overarching allegiance to the shared polity which counteracts these centrifugal forces. (Wilford and Wilson, 2001: 6)

Such an outcome was not to be unexpected given that devolution, regionalisation and even the evolution of postmodern politics can each unravel the complexities of ethnic belonging. Pratt's (2000) work on the disintegration of Yugoslavia clearly indicates how the politics of accommodation and consensus-building can run up against the bulwark and complexities of ethnic belonging.

The power of memory, perceptions of victimhood and the sense among unionists in particular of cultural dissipation have undermined the development of a political settlement based upon vague aspirations and a complete failure to analyse the structure and emotiveness of ethno-sectarian affiliation. In certain ways, the Agreement validated the importance of competing identities through wrongly acknowledging ethno-sectarian blocs as benign 'traditions'. The masking of ethno-sectarian competition as 'tradition' presented alternative and invalid cultural positions as both feasible and correct. Ethno-sectarianism was also to be achieved through the eloquent nonsense of 'parity of esteem'. If anything, the Agreement was not about challenging the 'comforting illusions about themselves that each ethno-sectarian bloc clings to with moral indignation' (Aughey, 2005: 15). As argued by Aughey, post-Agreement politics remained trapped within 'a mythological imagination which tends either to overestimate the potential of one's own side – an exercise in hubris – or to overemphasise the demonic potency of the other side – an exercise in paranoia' (2005: 26).

Within the Northern Irish context it is evident that devolution could not resolve political antagonisms that are rooted in the perpetuation of partition, armed paramilitary groups and the territorialisation of wider cultural, social and economic claims (Nairn, 2001; O'Leary, 1999; Stewart and Shirlow, 1999). It is important to understand that disagreements between the pro-British and pro-Irish populations remain and that devolution had a multiplicity of political and cultural meanings.

Determining the incapacity of Northern Irish society to shift towards pluralist and less culturally subjective categorisations of belonging and political devotion is of crucial importance. There has been no post-nationalist turn that has led to a decline in the link between national identity and electoral outcomes. Arguments concerning the rise of a post-nationalist Ireland are based upon a fundamental failure to recognise the perpetual power and passion of identity construction within a divided political landscape. There were in effect two evident flaws within the Agreement. First, it did not challenge the nature or position of ethno-sectarianism. Second, the return of a devolved administration did not resolve the long-term future of Northern Ireland's place within the United Kingdom. As prophetically argued by Nairn: 'A holding operation may have been undertaken in Northern Ireland; but although this is working for the moment, it is unlikely to last' (2001: 55).

In essence the Agreement's main success was what it partly achieved. It partly encouraged the political parties to deal with each other in a more inclusive manner. It partly gave shelter to wearied voices that had shifted away from the militancy of armed violence. It partly indicated that power-sharing was no longer denied by unionists on simply sectarian grounds, and it partly indicated the possibility of an emergent civic republicanism. The Agreement shifted the presentation and volume of ethno-sectarian competition but did not challenge the basis upon which it was founded and reproduced.

THE NORTHERN IRELAND ASSEMBLY

The problem for any writer on the Northern Ireland Assembly (NIA) is to choose the correct tense in which to write. The formations, collapses and reformations of devolution in Northern Ireland make the task of grammatical correctness increasingly difficult.

The Agreement was based around a series of political arrangements, which endorsed a power-sharing Executive, which, in turn, administered government departments via principles of proportionality. The aim of the Executive was not only to administer certain government departments but to also promote cultural and political equality. Voting within the Assembly was not determined by weighted or simple majorities but by parallel procedures of intercommunity consent. Key issues are only passed by the NIA if there is an overall majority plus a majority of both unionists and nationalists present at the time of voting. In order to uphold this intercommunity dimension each member of the NIA designates themselves as 'nationalist', 'unionist' or 'other'. It is possible to change designation, as was the case when members of the Alliance Party registered themselves as 'unionist' in order to re-elect David Trimble as First Minister after it became clear that he could not gain a majority unionist vote. Despite nominating themselves in such a way the Alliance Party, as well as other commentators, have identified these parallel procedures as institutionalising sectarianism. As contended by the Alliance Party:

But rather than these communal divisions being addressed and overcome, they are becoming institutionalised. The dominant orthodoxy is that separate but equal communities can be managed through some form of 'benign Apartheid'. However, skilful conflict management cannot be constantly maintained. With little or no common bonds or overarching loyalties to a set of shared values, once there is a major crisis, it is relatively easy for separate communities to go their separate ways. (no date)

The binational nature of the Agreement is also endorsed in that NIA members are not required to undertake an 'oath of allegiance' to the Crown and the British Union. Instead a 'pledge of office' requires support for exclusively peaceful means, democratic politics and support for the NIA and its Executive. The Northern Ireland Office, which is the Office of the Secretary of State for Northern Ireland, retains responsibility for constitutional and security issues as they relate to Northern Ireland.

At the centre of this consociational arrangement is Executive power-sharing. The 108-seat NIA has a dualistic leadership with a First Minister (formerly David Trimble, leader of the Ulster Unionist Party (UUP)) and a Deputy First Minister (formerly Seamus Mallon and later Mark Durkan, deputy leader and eventual leader of the

Social Democratic and Labour Party (SDLP)). There is virtually no difference in the task and duties of these ministers. A central goal of the NIA was the removal of 'direct rule' from Westminster through the placing of high-level governmental competency for education, health, economy, social services, environment and finance in the hands of the Executive. The Agreement also aimed to:

- bring forth the decommissioning of paramilitary weapons and other forms of demilitarisation;
- reform the Royal Ulster Constabulary, especially in relation to encouraging Catholic membership;
- provide for equality provision and the endorsement of British and Irish cultural rights;
- endorse the principle of consent and the upholding of the regulation that Northern Ireland's constitutional status within the UK cannot be altered without majority support within Northern Ireland and the Republic of Ireland.

The Agreement was also relatively unusual in that it was endorsed, in May 1998, by international referenda in Northern Ireland and the Republic of Ireland. Around 71 per cent of voters in Northern Ireland and 94 per cent in the Republic supported the Agreement. Somewhat peculiarly, the vote was not broken down by electoral ward as is commonplace in elections. The suspension of this practice was undertaken owing to fears that any such spatial breakdown would indicate that the majority of unionists had voted against the Agreement. At best 51 per cent of the unionist electorate supported the Agreement. It was also assumed that a vote breakdown would have indicated that there was a small but significant rejectionist vote within certain republican-dominated communities.

In upholding the principle of consent, the 1920 Government of Ireland Act was modified and Articles 2 and 3 of the Republic of Ireland's constitution were altered. With regard to the Irish constitution, the reunification of Ireland became aspirational. The significance of dual referenda was that they signalled a more flexible interpretation of Irish and British sovereignty and in so doing upheld a theoretical form of joint sovereignty, a North–South ministerial council and embedded forms of cross-border cooperation. This reimagining of constitutional boundaries was based upon upholding both unionist majoritarianism and a significant Irish dimension in Northern Irish affairs. Despite the maintenance of unionist electoral dominance it was evident that the Agreement moved beyond the 'British' constitutional system through the advancement of an alternative and renewable constitutional settlement. For Nairn the nature of constitutional renewal was unwelcome within unionism:

That is, it [the Agreement] does not depend on formal conventions, understandings among chaps and reinforcing the mythic Sovereignty of the UK Crown. On the contrary, it inclines towards republican formality and modern constitutionalism. This is the basic reason why Unionists profoundly mistrust it. (2001: 59)

In economic and cultural terms the endorsement of equalisation between nationalist and unionist communities was held to be important with regard to breaking down ethnically defined labour markets and issues concerning other social and cultural inequalities. In terms of 'conflict resolution' the intended aim of devolution was to democratise Northern Irish society and to recognise that the issue of sovereignty could be addressed at a later date, when there was the electoral capacity to deliver an alternative constitutional settlement. Furthermore, what may be seen as incongruous processes, such as the diminution of British sovereignty via the association of the Irish state in Northern Ireland's affairs, or safeguarding the Union via the principles of consent, testify to a constitutional stratagem aimed at encouraging issues of sovereignty to become increasingly obscure. Of course the Agreement did not silence the meaning of constitutional desire. If anything the Agreement became a site within which the differences between constitutionally committed ethnic blocs were to be played out. The NIA produced an arena within which the volume of distrust and suspicion, between ethno-sectarian blocs, was to be publicly raised.

Somewhat ironically, the decline in violence that accompanied the peace process was paralleled by an intensified process of political opposition, inquiry and demand. A fundamental problem with the overall interpretation of peace-building was that it became fixated upon the instruments of governance as opposed to questioning the underlying structure of Northern Irish society, and the reality that if one side adopted the new structures the other side would turn the other way. Unionists in particular could not identify obvious benefits when changes were perceived as eroding unionist values and principles. For republicans in particular, gaining political equality was a meaningful goal.

For both communities the politics of constitutional reformulation was less important, especially with regard to electoral outcomes, than the day-to-day politics of identity and political representation. Equality, for example, whether attached to housing, marching or employment, remained centred on a dominant politics of sectarian headcounting and the use of the electorates attached to sectarianised space.

AIMS AND OBJECTIVES OF THE AGREEMENT WITH TWO NAMES

The aim of the Agreement was to draw together atavistic political groups in order to promote a consociational accord, which would endorse Northern Ireland's place in the UK but at the same time uphold community rights and cultural expectations (Lijphart, 1999). The Irish and British states remain convinced that the genius of 'conflict resolution' lies in the competence of both states to fashion institutions that reconcile order with personal, spatial and communal liberty. Given the ethno-sectarianised nature of Northern Irish society, a central attribute of 'conflict resolution' was the need to engender the illusion whereby 'each side' would gain goals and objectives without being seen to lose out with regard to the 'other' community's beliefs and interpretations.

This process of confection overlooked the actuality that the experiences attached to the development of the Agreement would remain contested given that the ownership and presentation of oppositional discourses remained the basis of political motivation. As evidenced, the playing out of consocialisationist principles provided an arena within which the binary of resource competition became increasingly evident. Debates, for example, concerning discrimination were no longer framed around the relative position of the Catholic population with regard to that of Protestants. What began to emerge, in a more forceful and organised way, was the perception that 'concessions' to the minority population had ushered in an unequal treatment of the majority community.

This was illustrated by the decision of Sinn Fein's Bairbre de Brun, as Health Minister, to close the Jubilee Maternity Hospital (in a predominantly Protestant area) in order to upgrade maternity facilities at the Royal Victoria Hospital (located in a republican area). The closure of the Jubilee was viewed by many unionists as a purposeful act that denied Protestant access and ultimately undermined their 'right' to locate facilities in safe environments. The financial and organisational reasons that explained this transfer of facilities were obscured by more 'obvious' ethno-sectarian explanations.

This failure to move beyond ethno-sectarian issues of resource competition was also linked to the location of mobile capital investment, the siting of new road systems and issues relating to the opening and closure of schools. On one risible occasion a unionist politician, in a debate regarding financial impropriety within farming, argued that fraud was most likely among sheep farmers, who were predominantly Catholic. In refuting this, a SF representative alleged that cattle farmers were predominantly Protestant and that fraud was more commonplace among them than within sheep farming. Such debates remained far from what was being sought via political dialogue.

However, the actuality of enduring conflict was closely tied to the reality that the Agreement was underwritten by a desire to manage as opposed to resolve pre-existing ethno-sectarian traditions and conflicts between them. The central goal of the two states was to be seen to sponsor 'parity of esteem' and 'mutual consent' via the sponsorship of political structures that underlined pluralism and the eradication of economic and cultural sectarianism (McGarry and O'Leary, 1999). A crucial feature of an expected political transformation was the pluralist notion that the right to promote Irish and British cultural traditions had been endorsed constitutionally. However, such arguments failed to remember that the most passionate renditions of identity in Northern Ireland are not merely cognitive, nor are they heavily influenced by the 'holding operation' of constitutional change that has taken place. These atavistic renditions are based upon material realities such as segregation, notions of territorial loss, an inability to share space within a host of urban and rural arenas and the experience of harm.

Identity formation and the undeniable burden that segregation placed upon Northern Irish society were neither determined nor considered within the Agreement. Moreover, the depth and significance of ethno-sectarianism, in whatever form, were only discussed within the NIA in a vague and non-compelling way. The most enduring spatial aspect of division was left unaccounted for in a manner that suggested that the main political parties had no desire or interest in pursuing the desegregation of society. There was little or no surprise that interface violence became an ever-increasing feature within segregated places given that political leaders played upon group identity and their respective 'mandates' in a manner that qualified the suspicions and mistrust that were commonplace within such communities. The bitterness and acrimony of politics within the NIA and the associated crises and instability did little more than to encourage resource competition, especially when the Agreement fortified the primacy of group rights. The Agreement's fundamental flaw was that it placed traditions and group equality before the higher and more dignified principle of individual rights. In political terms the logic of the Agreement was based on the centrality of 'tradition', and as a result of this the diversity of tradition and the ambiguity of identity were slowly eroded by the requirement to hunker down behind group identities that would protect and insulate group beliefs. In certain ways the Agreement encouraged ethno-sectarian separation.

A more hopeful element in the search for political stability has been recognition among certain paramilitary organisations that long-term military campaigns were both unsustainable and futile. If anything the peace process has delivered a recognition that politics must be shifted away from mythic ends to more reasonable means, and political discourse is to be sought through inclusive politics and non-violent strategies.

There is also an increasingly healthy attitude towards power-sharing within unionism and unlike the doomed Sunningdale initiative there was no significant section of loyalism that was prepared to organise against the Agreement in a violent and confrontational manner. The decline in political violence also suggests that republicans have shifted from a form of militant purity to a more appreciative and complex interpretation of Northern Irish society. If anything the demand for paramilitary decommissioning has gathered force while the demand for British withdrawal has shifted in terms of emphasis. Further shifts include a recognition by republicans that there is potential for reform in Northern Ireland and that devolution and local control are more acceptable than Direct Rule.

It would be boorish to deny the level and direction of political morphology in Ireland toward the end of the last century (Stewart and Shirlow, 1999). The decline in extensive violence, a cross-border referendum and the appearance at times of cross-party intercession testify to the development of quasi-constitutional politics. However, there is still a need to understand that the perpetual realities of political intransigence and cultural contestation that surround job allocation, marching, decommissioning, flag bearing and policing are ever present.

THE BREAKDOWNS

The experience of devolution and the repeated collapse of the NIA around allegations of spying, a failure of paramilitaries to decommission and a demand for a speeding up of demilitarisation have undermined faith in the Agreement. Opinion polls have suggested that the unionist community remains supportive of power-sharing but that they are strongly opposed to any process that does not lead to the decommissioning of paramilitary weapons. Furthermore, in longitudinal surveys there was a shared intercommunity sense that the equality agenda did not deliver significant benefits to the unionist community.

To a certain extent globalisation and a more unified Europe have meant that increased interstate harmonisation of social and economic policy appears less politically sensitive and contentious than it once was (Michie and Sheehan, 1998). There is nevertheless a critical problem in the idea of creeping porosity with regard to 'the Border' since it assumes a collective loss of memory regarding the meaning of a cultural and political construct that has been the basis of armed conflict. The expected outcome is that the newly established cross-border institutions will operate in such a way as to make the pro-British population less insecure, given the limitations of these institutions and pro-Irish sections clamouring less for radical constitutional reformation. Of course in class terms it was expected that the 'Protestant' middle and business classes in particular would find benefits in trading and cooperating with an Irish state that now fits the model of a pluralist and progressive society (Shirlow, 2001).

As far as the British and Irish states were concerned, one of the hoped-for outcomes of the Agreement was that it would promote a post-nationalist interpretation of identity on the island of Ireland (Eagleton, 1999). The Agreement contradicted this and created a forum within which strident demands from either 'side' were met with a series of blocks and impediments. The Agreement in a sense went through two phases. In the beginning there were signs of compromise with regard to the issues regarding the constitution and power-sharing. These were followed by the formation of the NIA and the removal of the kind of 'behind closed doors' politics that characterised the birth of the Agreement.

Once the NIA was formed politicians were exposed to greater media and public scrutiny. At this point compromise was harder to deliver as the spectacle of such acts was on public view. What then emerged was a politics of 'not giving with one hand but trying to take with the other'. The constant need to refer back to each electorate for cognition highlighted the extremely tentative nature of the political process. The limited ethno-sectarian nature of such politics also meant that concessions were not interpreted as a recognition of goodwill and a valid interpretation of complex political processes. Instead, flexibility and pragmatic intercommunity politics were seen, especially within unionism, as group losses as opposed to mutual gains. The fundamental problem was that at some point within the Agreement, political parties were going to argue that they could give no more, or that giving was not reciprocated.

The critical problem affecting the implementation of the Agreement was that political horse-trading undermined the capacity to deliver on commitments and in so doing eroded mutual confidence and prevented the development of trust. Critically David Trimble's failure to achieve adequate progress on ending paramilitarism and the view of anti-Agreement unionists that concessions were not the means by which to gain PIRA disbandment did much to undermine the Agreement.

Two key issues had the capacity to undermine the political process – police reform and decommissioning/demilitarisation. The reform of the Royal Ulster Constabulary (RUC) was linked to the Patten Report, which recommended a series of reforms intended to increase 'Catholic' membership of this organisation from around 8 to 50 per cent. The SDLP has joined the Policing Boards that were designed to push through the reform process. The RUC has now become the Police Service of Northern Ireland (PSNI), and with that change has come a new uniform, new and more pluralist symbols and positive discrimination towards recruits from a Catholic background.

For SF the Patten Report was diluted and police reform was limited because the British state 'bowed' under unionist pressure. According to SF the new police service remains partial, unrepresentative, unfree from partisan political control and beyond a proper human rights culture. SF also allege that police reform did not provide for a common ownership of policing, especially with regard to the inclusion of former PIRA members. SF further maintain that reforms did not remove a unionist ethos and related emblems, and that the maintenance of the reserve units and Special Branch indicated a lack of meaningful change. Moreover, it is argued by SF that policing lacks democratic accountability and that PSNI includes 'torturers and abusers of human rights'. Ultimately, the failure to 'deliver' what SF recognises as a new policing strategy has meant that they have refused to take seats on the Policing Boards. SF argue that this failure is due to the British government not delivering the promises of the Good Friday Agreement and the recommendations of Patten.

For unionists, especially those attached to the DUP, policing reform was based upon unreasonable concessions to republicans and nationalists. The reforms that took place were seen to demoralise a police force that had 'stood in the way of terrorism', despite allegations of collusion with loyalist paramilitary groups. The removal of the term Royal and the crowned harp from the police badge created a sense among many unionists that their cultural and symbolic identities were being denied and removed from public service. For members of the unionist community such changes aimed to present the RUC as blamed and unaccountable.

More crucially the folk memory of loss while on service among police officers had always struck a distinct and emphatic chord within the unionist psyche. The DUP, in particular, mobilised this sense of loss and used political rhetoric to move the debate away from reform and accountability and towards the ownership of loss and memory. Evidently, policing reform did little to assuage republican senses of democratic accountability but did much to bolster the anti-Agreement unionist bloc, which identified change as being based upon appeasement with perceived and irreversible unionist costs. Policing became the first of many issues that delivered little to any one group but did enough to uphold vociferous critiques of the Agreement by both republicans and unionists.

The most significant indicator of perpetual political conflict during the lifetime of the Agreement was the issue of decommissioning. It is this issue that did most to destabilise the NIA and then, eventually, pull it apart. According to SF, who were constantly pressurised over PIRA weaponry, the Agreement suggested that there was no clause, subordinate or otherwise, linking PIRA weapons decommissioning to SF's participation in the Executive. SF's objections and those of the other political parties are upheld by the vagueness of the section on decommissioning within the Agreement, which basically conceptualised the issue through the imprecise concept of influence. The Agreement committed parties to: 'Continue to work constructively and in good faith with the Independent Commission, and to use any influence they may have, to achieve the decommissioning of all paramilitary weapons' (The Belfast Agreement: Section 7).

The words 'constructively' and 'in good faith' are sufficiently formless to allow SF 'off the hook' but clear enough for unionists to argue that SF must persuade the PIRA to hand in weapons. Despite the inspection of certain PIRA arms dumps, in 2000 and 2001, by the Independent International Commission on Decommissioning (IICD) it is still the case that unionists wish to see weapons put beyond their, and not the PIRA's, interpretation of 'verifiable' use. Even though Trimble argued that the 'organised campaign (of PIRA violence) has gone because it was beaten' and that unionists could accept the conclusions of the IICD that 'weapons and explosives cannot be used without our detection' (interview on Radio 4, 21 December 2000) it was obvious after a series of other 'concessions' that the unionist community required both decommissioning and the disbandment of the PIRA. The eventual decommissioning of PIRA weaponry in 2005 was met by unionist claims that the process undertaken was beyond verification. By that stage Trimble's positive assertion that the PIRA had moved in a significant manner and that unionism had won was suggestive of a pyrrhic victory at best and a denial at worst.

The objections to SF's participation in the Assembly created a problem for the British state, which is keen to keep SF on board for fear that to overlook their demands will lead to increased republican opposition within (and outside) the PIRA. For the DUP in particular the ministerial positions held by SF were unacceptable. Despite some signs of political maturity, such as DUP members sharing platforms with SF politicians, it is evident that the discourse of belligerent unionism remains tied to constantly 'exposing' the link between the PIRA and SF. Within such a logic of 'determined' linkage, the DUP began an assault upon the UUP based upon Trimble allowing republicans into the Executive on three occasions without the decommissioning of weaponry.

The DUP also played upon unionist fears that the PIRA was still in existence through pointing to the 'realities' of PIRA gunrunning in Florida, the case of the Colombia three, the Stormont spy ring, the perpetuation of PIRA violence and the Northern Bank robbery. The PIRA claimed that such allegations were unfounded. The only sign of hope was that politics had shifted towards a 'who done it' of criminality as opposed to a process where paramilitaries engaged in acts that they 'legitimised' through claiming the enactment of violent and criminal actions. The important point was that the less paramilitaries claimed involvement in such acts the more it meant that they supported a wider peace strategy. The moment that they begin to claim such acts will be the time when we will know that the peace process has unravelled. Such subtlety was lost upon the DUP, which could not interpret the nature of certain changes despite evident teething problems with regard to a new direction for paramilitary acts.

As the Agreement lurched from crisis to crisis the DUP became insistent upon the removal of SF from the Executive and the standing down of the PIRA. Decommissioning had to be complete, timetabled, verified, transparent, witnessed and photographed. Demands from the DUP stretched beyond the wooliness of the Agreement's version of decommissioning events. There would also, according to DUP, be a long-term assessment of paramilitary activity before SF would be allowed to rejoin the Executive. When the PIRA did, according to the IICD, put their weaponry beyond use the DUP scoffed and claimed that their actions and their verification were both inadequate.

This polarisation of politics was not expected among commentators such as McGarry and O'Leary, who both championed the notion that the outcome of consensus-building would be to bring out the benign characteristics of rival identities and in so doing marginalise ethnic chauvinists more effectively. O'Leary even went so far as to predict that: 'Sinn Fein and the DUP would do very well under fresh elections only if they run on moderated platforms' (2001: 18).

As has been indicated by the 125 per cent rise in voters for the DUP, between the 1997 and 2005 Westminster elections (Table 2.1), the rejection of certain aspects of the Agreement has grown with both fervour and intent and there have been few signs that moderation has won the unionist day. For the DUP political culture is tied to an unswerving commitment to an ethnically defined conception of territorial sovereignty. As such, the halving of the vote for the UUP is part of a process within which many pro-Agreement unionist voters have either chosen not to vote or moved to a more rejectionist unionist position. The 37.5 per cent growth in votes for SF also indicates that electoral choices are increasingly being based upon the deliverance of a defined cultural and political identity.

The difficulty for Sinn Fein was that the new art of governance could not be seen to have eclipsed defined notions of sovereignty. Thus they have won political support through the contention that the Agreement will lead to a united Ireland and that the demands that they make, upon the two states and unionism, are tied to the genuine voice of Irish nationalism. Such politics have been read as a championing of the republican/nationalist cause without compromise or evident ideological shifts. The overall winners, in political terms, have been those who have exposed their ethnic credentials under the guise of rights and while doing so have disguised the concessions that they have made. The UUP and SDLP, who openly presented their concessions, have been slowly devoured by an electorate that has sought ethnically defined refuges.

Table 2.1   Westminster vote 1997 and 2005

Political party	Votes 1997	Votes 2005	% change
Democratic Unionist Party	107 348	241 856	+125
Sinn Fein	126 921	174 530	+37.5
Ulster Unionist Party	258 439	127 314	−50.7
Social Democratic and Labour Party	190 844	125 626	−34.7

Source: www.ark.ac.uk/elections/

The attempts at the end of 2004 to encourage the DUP to work with SF failed when it became clear that neither would allow the other to triumph through any form of demand recognition. The sterility of political atavism between the two main parties and the nature of finality in the demand made upon each other was articulated by Adams when he stated:

Ian Paisley delivered his 'acts of humiliation' speech. Mr. Paisley's desire to 'humiliate republicans'; to have republicans 'wear sack cloth and ashes'; and the DUP's constant use of offensive language, was not and is not the language of peace making. For many across nationalist and republican Ireland this was too much. Especially when the governments supported the DUP position that the PIRA be photographed putting their arms beyond use. (cited in the Guardian, 5 March 2005)

Governance strategies that were based upon constructive ambiguity were never capable of establishing ongoing negotiations, especially when it was evident that such ambiguity provided alternative perceptions and meanings. The Agreement established 'objectives' of intercommunity governance instead of direct legal guidelines and accordingly there remained multiple interpretations of what the means and ends actually entailed. For unionists decommissioning meant a short-term end to armed groups, but for republicans it meant a consultative and ongoing process. Northern Ireland's identities and their presentation were too embedded to get political interlocutors to change basic positions and interests. Even attempts to modify the rules, such as with regard to the Leeds Castle Talks in 2004, failed to encourage an abandonment of well-rehearsed stances.

The Agreement, as articulated by competing political discourses, was also undermined by a failure to harness a greater sense of stability and 'normality' on the ground. Even though there has been a significant shift from the violence of the 1970s and 1980s, events such as Drumcree and Holy Cross encouraged the need for identity-based hegemony within a political process that increasingly did not indicate an emancipation from ethno-sectarian discourse. Thus normality remained a disputed concept with regard to group-based wills and desires. The bolstering of the 'two traditions' model within the Agreement merely affirmed examples of irreconcilable positions, but in so doing did not stipulate which one of them was the correct political interpretation. In such a vacuum ethno-sectarian politics seeks to overcome ambiguity through the endorsement of group rights.

OTHER REALITIES AND VIOLENCE

Northern Ireland lacks a unified 'lifeworld', which impedes a communicative dimension that means more than merely managing intercommunity division. The Agreement may well have established contact between oppositional groups, but that contact is based upon managing disorder as opposed to removing the meaning of communal mistrust and division.

The key problem facing Northern Ireland is the impact of ethno-sectarianism upon labour markets and consumption relationships (Poole and Doherty, 1996). The narratives and reality of protecting place are themselves interlinked devices in the whole enactment and reproduction of political and cultural identity (Jarman, 1997). Continuously being remade, the manufacture of territorial separation summarises dissimilar, eulogised and communally devoted places, inhibited by their contrariety to the territorial 'other' (Shirlow and McGovern, 1998). High levels of religious/political segregation, most common in urbanised and working-class areas, mean that mobility between communities is restricted owing to concerns over security and a desire to avoid the 'other' community (Burton, 1978; Darby, 1986; Douglas and Shirlow, 1998).

In many instances the political instability that still exists reflects the limitations of the Agreement and the inability of devolution to substantially alter the nature of conflict. The central goal of the Irish and British states is to be seen to promote 'parity of esteem' and 'mutual consent' via the promotion of political structures that underline modes of pluralism that remove the realities of economic and cultural sectarianism, especially those modes articulated since partition by the Northern Ireland state. However, as evidenced by the information presented within this book, sectarian actions are still in certain arenas more capacious than justifiable and secular words.

Despite the cessation of most paramilitary violence we are left with a situation within which the creation of territorial division and rigidified ethno-sectarian communities means that fear and mistrust are still framed by a desire to create communal separation. Residential segregation still regulates ethno-sectarian animosity via complex spatial devices. The reproduction of violence and fear is still achieved through linking ethno-sectarian affiliation and residence through the spatial confinement of political and cultural identity. The narratives and reality of constantly protecting place and religious segregation are still interlinked strategies in the ratification of friction and tension. More importantly, the ability to reconstruct identity and political meaning, away from ethno-sectarianism, is undermined by political actors who mobilise group rights in order to reinforce unidimensional classifications of political belonging.

More crucially, community-based self-representation assumes the form of a mythic reiteration of purity and self-preservation. The potential to create intercommunity understandings of suspicion, in terms of politics, is marginalised by wider ethno-sectarian readings. Indeed, for political actors the capacity to win political support has been based upon delivering a singular narrative of victimhood and exclusion. To accept now that such political vocabularies and actions victimise the 'collective other' would be politically unwise.

Intracommunity divisions testify to the reality that conflict transformation is not merely destabilised by intercommunity-based sectarianism. Wider senses of powerlessness are responsible for the failure of intercommunity politics to emerge. Political divisions provide a sense of the localised form of territorial control and resistance, where the goal of communal and intracommunal difference, separation and rejection still preponderate the politics of shared welfare and community sharing.

The Agreement and associated approaches to conflict management have attempted to build a civil society that will deconstruct the regeneration of sectarianised identities through engendering non-sectarian social relationships. As such the institutional approach is tied to developing intercommunity sharing and acknowledgement. There can be no doubt that there has been increasing dialogue between community representatives and an attempt by some, including many former combatants, to challenge some of the many fractures within Northern Irish society.

But there are several evident difficulties with regard to building cooperation at the community level. First, many of the solutions designed are top-down and reduce the problem to one of job creation and the stimulation of enterprise and small business formation. It has been well established that job creation does not simply reduce the nature and meaning of identity. Second, funds have been supplied in order to support what is known as single-identity work. The aim of such work is to 'educate' communities about their own history and culture. As with the Agreement more generally, the 'cultural traditions' perspective has provided greater legitimacy to these manufactured identities. This is not to deny that some of this work has not created meaningful intercommunity contact, but it is evident that such work bolsters the notion of cultural separation and resource competition. It is difficult to examine where such an approach identifies non-sectarian theorising and the revision of monolithic notions of identity formation. It is also unclear where such a single-identity approach confronts a process where history is emancipated from internal community interests. Where meaningful work is undertaken with regard to discouraging criminality and membership of paramilitary organisations, funding is uneven and limited. As noted by a community activist and former loyalist prisoner from Shankill:

You get these forms that ask you about how many people are you going to train in IT or what qualifications do you have. I have never seen a form for a grant that asks about how many wee lads are you going to stop from killing Catholics or throwing pipe bombs. There is money there but it is not related to the world we work in. Young lads about here can hardly read and write but the funders want you to teach them things that they just can't do. It makes me wonder if people up there really want us to build the peace on the ground using the experience and trust that we hold.

A fundamental problem is that the use of the community sector is based upon a fire-fighting exercise that lacks clear emphasis on meaningful conflict transformation. It is difficult to determine if there is a clear understanding of the aims and objectives of community work that is based upon intercommunity contact. Furthermore, it is increasingly evident that in certain instances dialogue between interface communities has imploded under the stresses and strains of ongoing violence. A general and emergent sense is that the funding of conflict transformation is based upon presenting the appearance of power being devolved downwards through new mechanisms of accountability, without any immediate aims or objectives.

A further limitation is the interpretation that new environments of trust and dialogue will emerge between communities without any consideration that identity formation in Northern Ireland is based upon a series of in-built mechanisms that aim to deny the logic of the 'other' side. The success of a state management of community resources has been to shift resistance to the state into new processes of lobbying and grant provision. Thus the problems that communities face can, it is assumed, be resolved via institutionalised forms at the local level.

Such a process of management founders when ethno-sectarian disputes over marching and housing issues create disorder and draw communities into confrontration with the power bloc that is built around the state and the PSNI. We now live within an environment in which financial and political resources are localised, and many of those communities and individuals who receive such support are no longer considered as being both politically deviant or socially transgressive. In practical terms the state has shifted towards a more pragmatic relationship in which the community/voluntary sector is utilised as a circuit breaker of communal angst. The goal of the state is to provide facilities that allow the community/voluntary sector to police themselves. This has been a significant development in that, especially with regard to former combatants, a relationship has been established within which communal disorder can be solved within a local context.

However, as noted, the problem is that when ethno-sectarian issues arise the relationship with the state reverts to positions of open hostility. In many instances, the new relationship between the state and the community/voluntary sector is improving but ultimately it is somewhat schizophrenic. Despite more creative and managerial mechanisms within which to roll out conflict transformation it is evident that violence has remained a relative and constant feature of Northern Irish society since the ceasefires of 1994. A more common form of violence has been attached to a swathe of sectarian attacks upon symbolic sites such as churches, Orange Halls and GAA clubs.

According to Jarman (2005), there were 594 reported attacks upon these symbolic sites. Jarman also concluded that there were a staggering 6,623 sectarian incidents in Northern Belfast between 1996 and 2004.

There were 1,129 recorded punishment attacks since 1994, the year of the first paramilitary ceasefires (Table 2.2). As with more recent violence, the appearance of such attacks was rare between the ceasefires and the advent of the Agreement. There were virtually no punishment attacks in 1995 and relatively few between 1996 and 2000. The onset of instability following the three suspensions of the NIA between 2000 and 2004 was matched by a near doubling of such assaults. The attempts in 2004 to re-establish the NIA heralded a small decline in this particular form of violence. A common suggestion among unionists and the press is that such violence declines during periods of political negotiation and resumes when dialogue breaks down.

Table 2.2   Punishment attacks in Northern Ireland 1994–2004

Year	Assaults	% of all assaults 1994–2004
1994	122	10.8
1995	3	0.2
1996	24	2.1
1997	72	6.3
1998	72	6.3
1999	73	6.4
2000	136	12.0
2001	186	16.4
2002	173	15.3
2003	156	13.8
2004	112	9.9
Total	1129	100

Source: Authors' calculations

Two hundred and twenty-five persons were murdered between 1994 and early 2005 (Table 2.3). The significance of a declining death rate is evident given that in the previous ten years there were 857 deaths. This equals a 74 per cent decline in the number of deaths between these two periods. The nature of violence has also altered in that the death rate among those who are police officers or members of the prison service or related army personnel ceased in 1999.

Table 2.3   Politically motivated deaths in Northern Ireland 1994–April 2005

Year	Non-civilians	Civilians	Total
1994/95	5	49	54
1995/96	0	12	12
1996/97	2	12	14
1997/98	4	29	33
1998/99	2	42	44
1999/00	0	7	7
2000/01	0	18	18
2001/02	0	17	17
2002/03	0	15	15
2003/04	0	7	7
2004/05	0	4	4
Total	**13**	**212**	**225**

Source: Authors' calculations

In sum, 13 security-related personnel were killed in the more recent period compared to 272 in the preceding ten years (Table 2.4). Civilians constituted 67 per cent of all victims in the 1983 to 1993 period compared to 94 per cent in the period after 1994. In addition, the death rate among civilians had declined at a much slower rate (64 per cent) than was the case among security-related persons (95 per cent).

Table 2.4   Politically motivated deaths in Northern Ireland 1983–93

Year	Non-civilians	Civilians	Total
1983	33	44	77
1984	28	36	64
1985	29	26	55
1986	24	37	61
1987	27	68	95
1988	39	55	94
1989	23	39	62
1990	27	49	76
1991	19	75	94
1992	9	76	85
1993	14	70	84
Total	**272**	**575**	**857**

Source: Authors' calculations

The events around Drumcree in 1998 witnessed a doubling of the death rate located in the period between 1995 and early 1997. The high death rate in 1999 was heavily influenced by the Omagh Bombing. Many of the deaths that have occurred in the past decade have also been related to loyalist feuds, but there have also been ethno-sectarian murders in segregated communities in areas such as Rathcoole and between Ardoyne and Upper Ardoyne (Table 2.3).

From 1994 to 2004 there were 2,505 shooting incidents (Table 2.5). This is equivalent to around a half of all shooting incidents between 1983 and 1993. There is less variation with regard to shootings than is the case with other forms of violence, although 1995 witnessed a dramatic fall in such acts. There was some growth in shooting incidents around the time of the Holy Cross dispute and also around the instability caused by various political crises such as Drumcree. As shown in Table 2.6 the use of bombs fluctuated within the 1994 to 2004 period, and as with shooting incidents the number of such acts halved in comparison to the 1983 to 1993 period (Table 2.6).

Table 2.5   Shooting and bombing incidents 1994–2004

Year	Shootings	Bombing devices used
1994	348	222
1995	50	2
1996	125	25
1997	225	93
1998	211	243
1999	125	100
2000	302	135
2001	355	444
2002	350	239
2003	229	88
2004	185	69
Total	**2505**	**1660**

Source: Authors' calculations from data supplied by PSNI

Most of the bombing devices used were pipe bombs, which are less destructive than the majority of bombs used in previous violent periods. The use of pipe bombs has been largely undertaken by loyalist paramilitary groups within interface areas. Pipe bombs are rudimentary and small weapons that are essentially used as a hand grenade. Their use in rioting and the intimidation of families within interface communities became commonplace during and immediately after the Holy Cross dispute.

Table 2.6   Shooting and bombing incidents 1983–93

Year	Shootings	Bombing devices used
1983	424	367
1984	334	248
1985	238	215
1986	392	254
1987	674	384
1988	538	458
1989	566	420
1990	557	286
1991	499	368
1992	506	371
1993	476	289
Total	**5204**	**3660**

Source: Authors' calculations from data supplied by PSNI

Table 2.7 examines data on intimidation cases reported to the Northern Ireland Housing Executive as a result of civil disturbances between 1989–99 and 2000–1. Just over half (52 per cent) of all incidents took place in Belfast. One of the striking features of the data is the dramatic rise in the number of reported cases post-1998.

Table 2.7   Profile of intimidation cases by Housing Management District

District	98–99	99–00	00–01	Total	%
Belfast 1 West	3	21	34	58	3
Belfast 2 East	3	26	40	69	4
Belfast 3 West	3	32	36	71	4
Belfast 7 South	9	60	90	159	9
Belfast 4 North	1	40	104	145	8
Belfast 5 Shankill	1	11	238	250	15
Belfast 6 North	18	28	110	156	9
Total Belfast	41	218	652	908	52
Total Northern Ireland	**56**	**549**	**1108**	**1713**	**100**

Source: Data supplied by NIHE

In relation to housing intimidation within the Shankill district most of this intimidation was linked to intercommunity-based loyalist violence. Despite this the majority of intimidation has been against spatially vulnerable minority populations within distinct social housing arenas.

The rise in intimidation from as low as 56 cases in 1998/9 to 1,108 cases in 2000/1 highlights the nature and extent of violence in the wake of the environments created by political instability. It should also be noted that these figures do record those persons residing within the private sector who moved from their homes due to intimidation.

The obvious effect of such violence was that it eroded the goodwill experienced in the immediate ceasefire period, and at the same time created a new generation who experienced dramatic violent events within their lifetime. The problem with the Belfast Agreement was not merely that it remained unsustainable but that the insecurity and political mayhem that came about forged new versions of victimhood and risk at the hands of the ethno-sectarian 'other'. The complexity of the new political dispensation was adequately and forcefully summarised by the leftist commentator Eamonn McCann:

It's because some Nationalists are uneasy at their own acceptance of Northern Ireland that they feel they have to make a show of rhetorical opposition to it. It is because, in practical terms, they have endorsed the legitimacy of the Northern Ireland State that they denounce symbolic representations of it all the more loudly. The campaign to obliterate Northern Ireland having halted, they turn to battle on who'll rule the roost within it. Communal hostility replaces the struggle for an all-Ireland. This is a pattern of play which corresponds ever more closely with the political mind-set of the Mad Mullahs of Orangeism. (2005:21)

The Agreement posted the journey upon which solutions could be found. The building of new and more sophisticated techniques of conflict management was a positive development in the attempted renewal of Northern Irish society. However, the subtlety needed to shift society forward has been countered by a growth in the link between ethno-sectarianised identities and electoral outcomes. In plain terms, the politicians who did most to shore up pluralism, admittedly in a muddled way, floundered on the beach of ethno-sectarian competition.

3

Interfacing, Violence
and Wicked Problems

Residential segregation has been a prominent feature of urban division within Belfast since the onset of industrialisation in the nineteenth century. During periods of relative calm and political stability many segregated places had zones between them that were relatively mixed. Any rise in political tension or in sectarian violence, as in the 1920s, 1930s and more recently, tended to result in these mixed zones becoming sites of violent enactment and intensified border demarcation (Boal, 2000; Feldman, 1991).

The violence enacted since the late 1960s has essentially removed these residentially mixed zones and replaced them with uninhabited border zones. Unlike the sectarian violence of the 1920s and 1930s, more recent processes of interfacing have been linked to augmented and sustained modes of conflict and political discord. The subsequent increase in segregation since the 1960s has encouraged a series of bordering events that have furthered the meaning and significance of previous ethno-sectarian divisions. As argued by Feldman, the growth in urban interfacing has been 'in symbiosis with the pattern of sectarian residential extension' (1991: 28).

For many residents of segregated places the borders between unionist/loyalist and republican/nationalist spaces are not merely boundaries between communities but important instruments in the definition of discursively marked space. Interface walls are not interpreted as protective barriers that impede violent enactment. For some, they are crucial structures that reduce contact with a lifestyle and cultural designation that is culturally, politically and socially 'improper'. The immediate impact of interface walls is to create social, political and cultural distance between communities. The capacity of such boundaries to turn small-scale physical distances into expansive symbolic signs of cultural and political differentiation is both significant and undeniable.

Interfaces are also a constant reminder of harm done and of threat implied. Their existence compacts the performance of violence into space and they present a script from which community loyalty can be read. Interfaces both divorce and regulate intercommunity relationships, and in so doing they compress space into sites that become the most notable places of violence and resistance.

Interfaces, which in many instances are sites of militarised barricades, have over time become self-sustaining features. There has also been a demand for new interface walls between segregated communities during the current peace process. The loyalists, for example, who drove the Holy Cross dispute were only persuaded to cease their campaign once they were assured that an interface wall and other security measures were to be placed between them and the republican/nationalist population of Ardoyne. A senior Ulster Defence Association leader labelled this the 'pipe bombing, wall getting campaign'.

As noted the boundaries between segregated places are borders between competing ideological perspectives (Crowder, 2000; Squires and Kubrin, 2005). Within the Belfast context, most of these boundaries have been or remain as spaces of violent exchange within which communities still endure harsh and adverse conditions (Sayer, 2000). Such adversity has created a series of intracommunity exchanges within which the material practices of protection, sacrifice and loss at the hands of an ethno-sectarian 'other' encourages spatial enclosure and the colonisation of community beliefs. Communities have become not merely sites of residence but sanctuary spaces defined by complex internal and external forces. The spatialisation of fear, violent resistance and the desire to promote discourses of internal unity while under threat has redefined, rebuilt and delivered more impassioned forms of space-based identity. In short, walls have multiple and contradictory functions: they shape and are shaped by conflict and, as Throgmorton noted:

Not all intentional walls are alike. They can differ in terms of their purpose, transparency, permeability, mutability, construction materials, difficulty of passage, compatibility with their contexts and effects on the imagination. While walls separate *this* from *that*, for instance, all walls also link and connect.

They accomplish this by incorporating gateways, passages or bridges, and by providing sites of transition, such as meeting places, playgrounds, shared functions and shared experiences. Hence even the hardest of walls can also act as 'seams' or 'zippers' that knit opposing sides together. Moreover, the *meaning* of a wall is not an objective fact, independent of interpretation. (2004: 250–1)

SEGREGATED SPACES

It is difficult to gauge the exact level of segregation within Belfast for several reasons. First, segregation is relative and is understood in multifarious forms. Some places are overtly impacted upon by segregation and there are active practices within them that aim to maintain a form of community dominance. Mixed places are obviously less physically segregated but citizens living in shared environments may maintain a reliance upon institutions and places that are designed via ethno-sectarian choice. Notable examples, as shown in Chapter 5, include schools, sports clubs and other sites of leisure and consumption. In effect, mixing does not mean that the residents of such places integrate around the same series of cultural and social practices.

There is a general suggestion that as socially mobile Catholics move into mixed middle-income communities there is a tendency for some Protestants to move out. Whether there is any validity in such a notion it is clear that many private housing arenas within Belfast have been tending towards more significant shares of Catholics (Poole and Doherty, 1996). The desire to create mixed communities is not an important motivational factor; rather, such population shifts through increased mixing are primarily tied to burgeoning social mobility among nationalists and republicans within the city (Murtagh, 2002).

Second, the spatial units that are used to construct census boundaries are altered on each occasion that a census is undertaken. Therefore, it is impossible to link sets of data together to determine how and in what ways the religious demography of places has altered.

Third, census-based data may suggest that an output area is mixed even though it is evident that such units contain populations who live within different parts of the spatial unit being measured. The sundering of such units by interfaces has never been adequately determined and consequently the extent of segregation can be hidden within asocial units of reference.

Finally, the meaning of segregation is in some instances disguised, and it is the perception and meaning of segregation that is more profound than any mechanistic census-based measurement of that division.

As shown in Table 3.1, the majority of persons from a Catholic or Protestant community background live in places that are at least 81 per cent Catholic or Protestant. Just over two-thirds of Catholics (67.3

per cent) and 73 per cent of Protestants live in such places. A mere 10.7 per cent of Catholics and 7.0 per cent of Protestants live in places that are between 41 and 60 per cent Catholic or Protestant, places that could be described as mixed. There is a near equal population split between these two populations within the city, a situation that reflects an increasing share of the population emanating from a Catholic community background.

Table 3.1   Segregation in Belfast by community background

% Population bands	Community background % of total Catholic population in band	Community background % of total Protestant population in band
0–20	4.7	3.4
21–40	3.6	7.3
41–60	10.7	7.0
61–80	13.8	9.3
81–90	9.3	28.4
91–100	58.0	44.6

Source: Census of Population, 2001

A common source used in order to determine if segregation is increasing or decreasing is that of the Northern Ireland Housing Executive (NIHE). The NIHE consider housing areas as 'segregated' if over 90 per cent of residents, within each estate, are from a particular community background. Their research constantly confirms that around 98 per cent of NIHE (1999) estates in Belfast are segregated compared to a rate of 71 per cent for Northern Ireland. Their general conclusion is that segregation within social housing estates is rising.

As shown in Map 3.1, the mosaic of segregation within the city is complex given the combined issues of scale and interpretation. There are evidently large areas, such as West Belfast, the Shankill and extensive parts of East Belfast, that are dominated by one of the two main communities. Such large single-identity zones are places within which the interface between communities tends towards a historical lineage.

The intricate territorial divisions of North Belfast create some of the most complex geographies of division within the city, a reality that has been reflected in the performance of violence. Within North Belfast the borders between nationalist/republican and unionist/loyalist communities are constantly being altered by a series of demographic shifts. Unlike most parts of the city, North Belfast has a complex series of ever-changing interfaces and spatial juxtapositions, a process of change that is closely linked to the recent reproduction of violence in that part of the city.

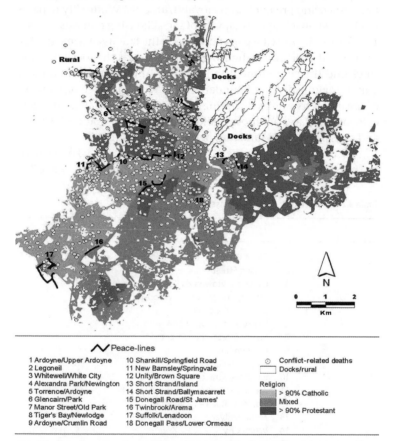

〜 Peace-lines		
1 Ardoyne/Upper Ardoyne	10 Shankill/Springfield Road	⊙ Conflict-related deaths
2 Legoneil	11 New Barnsley/Springvale	▢ Docks/rural
3 Whitewell/White City	12 Unity/Brown Square	Religion
4 Alexandra Park/Newington	13 Short Strand/Island	▮ > 90% Catholic
5 Torrence/Ardoyne	14 Short Strand/Ballymacarrett	▮ Mixed
6 Glencairn/Park	15 Donegall Road/St James'	▮ > 90% Protestant
7 Manor Street/Old Park	16 Twinbrook/Arema	
8 Tiger's Bay/Newlodge	17 Suffolk/Lenadoon	
9 Ardoyne/Crumlin Road	18 Donegall Pass/Lower Ormeau	

Map 3.1   Prominent interfaces, segregation and politically motivated deaths in Belfast 1969–October 2005

Source: Authors' calculations

Boundaries within North Belfast have tended to increase as mixed areas, such as the Antrim Road and Cliftonville Road have become more nationalist/republican. New interfaces have also been created between private housing areas and public sector housing estates. The interface between White City and Whitewell has become a relatively new arena within which ethno-sectarian violence is being played out. This growth in violence is directly linked to the private housing community around Whitewell being redefined as a predominantly nationalist/republican community while the established and predominantly loyalist/unionist White City remains unchanged owing to the majority of social housing that exists there. This form of increased segregation reflects the fluidity of demographic shifts within the private housing sector, whereas the social housing stock and the commitment to identity within such places present a more rigidified set of boundaries. The term interface has a series of meanings as well as operating at various spatial scales. The most evident interfaces are those marked by high walls that both sunder and demarcate the boundaries between communities .

A list of the current physical divisions in the city, as acknowledged by the Northern Ireland Housing Executive, is set out in the Table 3.2.

Table 3.2   Physical interfaces in Belfast at 2005

Interfaces
1.   Cluan Place
2.   Bryson Street
3.   Lower Newtownards Road
4.   Mountpotinger Road – Woodstock Link
5.   Duncairn Gardens
6.   Henry Street – West Link
7.   Manor Street – Roe Street
8.   Crumlin Road
9.   Alliance – Glenbryn
10. Elmgrove Street
11. Alexander Park
12. Hallidays Road – Newington
13. Mountcollyer
14. White City
15. Longlands
16. Mountainview
17. Squires Hill – Hazelbrook
18. Northumberland – Ardmoulin
19. Cupar Way
20. Ainsworth
21. Springmartin
22. Springhill
23. Suffolk
24. Roden Street
25. Carrick Hill – Peters Hill

It should be stressed that there are several lists of interfaces compiled by various agencies that differ, given the varying interpretation of what constitutes an interface.

Some interfaces are not demarcated in such an obvious manner. The interface on Skegoneill Avenue in North Belfast, for example, is demarcated by a series of derelict homes and flags that mark the cessation of nationalist/republican space and the beginning of loyalist/unionist territory. Other interfaces are simply understood through a knowledge of ethno-sectarian division, and can be marked by material forms such as a lamppost or bus stop. This is illustrated in Figure 3.1, which highlights the wider effects of the interface on the hinterland community as well as on those people living near the dividing line itself.

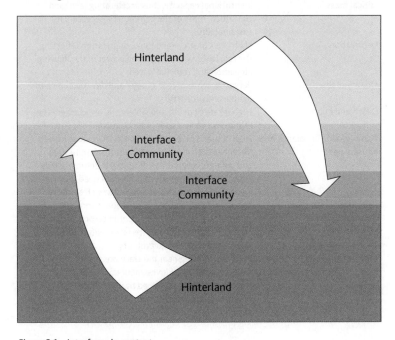

Figure 3.1   Interfaces in context

The Belfast Interface Project (2005), Shirlow *et al.* (2003) and Murtagh (2002) have all itemised the policy costs of living in an interface community. These are summarised in Table 3.3. Boal's (1969) seminal work on the Shankill–Falls Divide in Belfast showed how communities operated within their own territorial boundaries for shopping, services and communal and kinship interaction. Linked to this spatial dynamic is the issue of population mass and the institutions, facilities and services that a population can support. Poole and Doherty (1996) demonstrated a close connection between locality decline as a consequence of a community feeling uncertain about its ability to occupy a particular locality safely.

Table 3.3 Policy costs of interfacing

Issue	Description of need
Activity segregation	Activity segregation resulting in facilities and services 'trapped' in the territory of the out-group
Community institutions and critical mass	Locality population change undermines local institutional capacity, thus accelerating 'exit' and the critical mass of the community necessary for sustainability
Deprivation	High rates of socio-economic deprivation reflecting the residents' weak bargaining power in the housing allocation and transfer system
Quality of life	Pervasive sense of fear, danger and direct violence to people and property
Death and injury	Higher rate of death and violence in areas where the ethno-sectarian map is most contested
Demographic imbalance and housing need	High demand in republican/nationalist areas fuelled by higher than average fertility rates, family sizes and younger age profiles. Protestant demographics generate comparatively less housing need, particularly given the wider choice of housing search territory
Symmetrical land and property markets	The reproduction of segregated space through symmetrical and often self-contained property markets
Direct costs	Physical construction of interfaces, buffer zones and security adaptations to property
Blight of land and property	Land and housing near interface areas blighted by fear, violence and lack of investment confidence
Image	Negative imagery produced by walls of division, sectarian graffiti and physical dereliction to investors and tourists

We noted earlier that as 'exit' becomes a strategy for coping, the residing community is less able to support local facilities and services, including shops, schools and churches. An associated process is residualisation, whereby communities occupying marginal, dangerous or contested territory are increasingly characterised by high rates of social deprivation and poverty. The 25 physical interfaces in Belfast cover 22 wards, and of these, 17 (77 per cent) are in the top 10 per cent deprived wards as measured by the Noble Index[1] (2001). Added to this are the daily experiences of people living in an area affected by often low-level but constant violence, pervasive fear and the threat of attack. For example, a joint study by the Housing Executive and the Eastern Health and Social Services Board found that the residents of the interface areas experienced higher rates of long-standing illness, attendance at GPs and stress than the rest of the Northern Ireland population (NIHE and EHSSB, 1995).

The demographic characteristics of Protestants and Catholics also differ significantly in interface areas (Dunn and Morgan, 1994; Knox, 1995). Catholic communities are characterised by higher than average family sizes, higher fertility rates and a more youthful demographic structure. The Protestant population is older, with smaller household sizes and lower than average fertility rates (Compton, 1995). When these demographics are acted out in highly segregated space, very different housing need profiles result in terms of the number, size, type and location of accommodation required (Compton, 1995).

Paris *et al.* (1997) found evidence of symmetrical land and property markets in interface areas, with different sets of estate agents, developers and solicitors serving almost self-contained market systems that reproduce residential segregation in the private and rented housing market. The financial costs of interfaces are compounded by the implicit or explicit provision of dual facilities to serve different ethno-sectarian catchments. Much of the land and property within contentious areas is blighted for housing use, and the construction of buffer zones is not always possible given the configuration of the ethno-sectarian map in certain places.

The negative image that brutal lines of division, physical dereliction and poverty project to a wider audience can also be a major obstacle to investment and tourism (Neill, 1995, 1999). These policy costs are outlined in Table 3.3. As shown, the problems caused by interfacing stretch from the monetary to the symbolic. It is also important to note forms of masked interfacing that are based upon hidden forms of separation. In some instances, avoidance tactics are a benign form of removing the capacity for confrontation or respecting the dissimilar beliefs of those spoken to. However, outward friendliness towards residents from the 'other' ethno-sectarian group can also be accompanied by disguised emotions based upon mistrust and even concealed loathing. These latter behaviours are founded upon masking true emotions, which betray deeper senses of ethno-sectarian angst.

Interfaces may vary with regard to form, style, meaning and both visibility and invisibility. However, they will most certainly be known and understood by those who live within segregated communities. This knowledge of spatial demarcation is based upon territorial marking and the twin processes of telling and spatial identification. Such processes are a common feature for many who have grown up within the city. The marking of space is thus reinforced by a sectarianised and location-driven knowledge. The telling of division and the 'need' for separation are also founded upon a system of signs through which ethno-sectarian belonging is determined within both obvious and discrete spatial settings (Burton, 1978).

Segregation and related mobility issues mean that whole swathes of the city are virtually unknown to citizens living within Belfast. Persons of a pensionable age seem to possess a more collective and knowledgeable geography of the city. Such knowledge pre-dates the severe sundering of places that preceded the late 1960s. This does not mean that such persons were not aware of boundaries between communities, but these interfaces, prior to the 1970s, were easily crossed and at times traversed with enthusiasm. Similarly, for older age groups there have been more evident forms of intercommunity engagement, especially with regard to sharing space. As noted by a respondent, aged 68, from the predominantly republican New Lodge area:

Do you see out there in the street? That is where we all built the bonfire for the 11th night. It was also the place where we built another bonfire to celebrate Our Lady's Day on the 15th August.[2] There was (sic) never any problems living and sharing together.

As shown in Chapter 4, younger generations tend towards a more limited knowledge of places within the city. Evidently growing up within an intensively segregated environment, within which boundaries between Catholics and Protestants were rigidified by violence, has a consequential impact upon the interpretation of place. Age and ultimately the experience of segregation influence the form, conception and interpretation that shapes various geographies of separation and segregation.

IT WILL BE OVER BY CHRISTMAS!

The walls that were built between communities were in the first instance a reaction to the societal breakdown that accompanied the initial phases of violence between the late 1960s and early 1970s. In most instances, the British Army and vigilante groups erected barbed wire fences and other temporary instruments that served to reduce the capacity of communities to assault each other. It was initially conceived that violence would be short-lived, as had been the case with other rises in ethno-sectarian tensions. As a resident living along the Falls–Springfield Road interface noted:

I went up to one of the soldiers when all these troubles started. One of the top men, you know. I went up to him and asked him when them there (sic) fences were to be taken down and when we could all get back to a normal way of getting on.

'Sir, it will be over by Christmas,' he said. I should have asked him what Christmas he was meaning!

As the conflict bedded down into a more sustained mode of violence the erection of the oxymoronically entitled 'peace-lines' was undertaken with permanent intent through the use of brick and steel. The permanency of such constructions has been confirmed by many walls being heightened and reinforced in recent times. New walls have also appeared in recent years. Alexandra Park, in North Belfast, had a wall built through it in 1998. Somewhat ominously, the construction of this particular wall began during the first week that the Northern Ireland Assembly met.

Other physical barriers that demarcate republican/nationalist and unionist/loyalist places include main roads, industrial buffer zones (between Tiger's Bay and New Lodge) and areas of parkland. Barriers that close roads during periods of tension and gates that block entries are also commonplace. There has also been an attempt through landscaping and design to create a form of invisibility regarding such walls that might reduce the significance of a divided urban landscape. Signs of segregation are also denoted by the existence of grills and bars used to protect property. In addition, many homes along interfaces have security doors fitted as well as hosting reinforced glass.

(RE)MARKED SPACE

As previously established a central problem within Belfast is the link between territoriality and the political organisation of space. The work of Gottman (1973) within other segregated arenas contends that space is constantly partitioned through a combination of the physical and the symbolic. As articulated by Sack, the processes that constitute territoriality are primarily expressions of 'social power' (1986: 5). In this sense, as argued by Foucault (1980), power emanates from localised sources of transaction and politicised action. Such power, especially within the context of intercommunity violence, is not only destructive but also, in terms of action, both fluid and enabling.

The use of wall murals is a coherent form of territorial marking and an illustration of territorialised power in particular. The choice of certain events and images is tied to the presentation of 'dominant' senses of territorial belonging as well as expressing what Lefebvre (1995) would have interpreted as sites of active struggle, and political and cultural resistance. The marking of space in such circumstances pinpoints political selection through reminding the viewer of oppression and the need to celebrate examples of armed and civil resistance. Such markings span a spectrum from what Graham (1998) identifies as the mnemonic to the reinforcement of cultural apartness.

In particular, the representation of a spatially proximate event upholds the link between place and experience. Muralists use defined spaces within which to operationalise a propaganda-conditioning perspective that encompasses signals of territorial demarcation. The British state's use of 'anti-terrorist' advertisements on television, in a bid to censure competing ideological perspectives, further highlighted the use of imagery in the struggle for legitimacy and political control (Storey, 2001).

More immediate experiences have strong symbolic codes and messages attached to them. The departure in 2005 of the remaining Protestant community from the Torrens Estate in North Belfast was a signifier of the incapacity to solve evident interface difficulties. The leaving of the last few residents was viewed through disparate forms of analysis. For the residents the impact of republican violence and intimidation had been the main catalyst that encouraged their departure. For republicans, the exit of the Protestant community had been negotiated and welcomed by those who had left. The hoisting of Irish national flags moments after the departure of these residents presented a form of space-claiming that intensified the loyalist/unionist sense of decline and forced exodus.

The exit of this community was upheld within sections of unionism/loyalism as an example of 'ethnic cleansing'. Such unverified claims of ethnic cleansing are common within a unionist politics articulated around discourses of fear and ultimate defeat. As summarised within the following quote by the Ulster Protestant Movement for Justice the sense, although unsubstantiatied, of demographic decline is a crucial component in unionist/loyalist suspicions: 'An estimated 250,000 Protestants have been forced out of various areas of Northern Ireland … one in four of the Protestant community has moved under direct or indirect threats or intimidation' (www.upmj.co.uk/displacement.htm).

Torrens illustrated a fundamental problem with regard to the city's future and the obvious meaning of marked space. There are several predominantly Protestant communities that have undergone long-term demographic decline. Communities such as Oldpark and parts of Ballymacarrett now contain significant numbers of vacated properties. These places are unable to replenish housing tenancy, especially when the housing stock is located close to interface walls. For republicans in particular the existence of vacated properties should be resolved through the movement of these interface walls to accommodate housing need.

For loyalists and unionists any loss of territory is read as an example of state-based duplicity through the rewarding of Irish republicans in order to bolster the 'peace process'. Housing demand thus remains framed within an ethno-sectarian logic of community-based ownership of publicly funded space. Within Belfast, public housing space is defined through what may be perceived as ethno-sectarian control. The key issue is whether the state is incapable of controlling public resources or whether the appearance of community control is imagined. As noted by Jarman and O'Halloran:

The strength of feeling towards local territorial identity and boundaries is such that government and statutory bodies have generally, if unofficially, accepted that it is not yet possible to engage in any social geographic engineering to move peace-lines in order to build new houses or to reduce the potential for disorder. (2001: 2)

The key issues in terms of interpreting the meaning and nature of interfaces are those of memory, dissimilar demographic profiles and alternative perceptions concerning the future of housing need. Evidently, there is a strong perception within certain loyalist/unionist places of terminal spatial and cultural decline. As explained, examples of decline are upheld as being tied to the impact of republican assault upon Protestant places. Such presentations of Protestant anguish rarely consider the ongoing decline of the Catholic population in places such as Ahoghill, Bushmills, Carrickfergus and Larne.

Other important factors that encourage territorial decline within Protestant communities, such as social mobility, ageing, loyalist paramilitarianism and de-industrialisation, are rarely considered as factors that have, undoubtedly, impacted upon such losses. For republicans the presentation of their housing needs is rarely influenced by an introspective acceptance that their demands are seen as expansionist. In both instances communities have marked space and any dilution or expansion of such space is ultimately a politically divisive act.

TOWARDS ETHNO-SECTARIAN ENCLAVING

Until the recent Balkan conflicts, population movement in Belfast, resulting from intimidation, was the most significant shift of people attributed to violence within Europe since the conclusion of World War II. The influx of refugees (around 7,500 families in Belfast in the period 1968–2001) into ethno-sectarian enclaves, following the reappearance of violent conflict, created sanctuary spaces that functioned as safe and unsafe mental maps (Belfast Interface Project, 1999; Boal and Murray, 1977; Burton, 1978; Darby, 1986; Feldman, 1991; Poole and Doherty, 1996). The realities of politically motivated violence in Belfast are obvious. Between 1969 and 1999, around 1,400 people were killed and over 20,000 injured by paramilitary and state violence. Fear of being a victim of such attacks meant that many people living in conflictual arenas developed a comprehensive knowledge of 'safe' and 'unsafe places' (Burton, 1978). Furthermore, mental maps created a process of consciousness that was intensified through the telling of fear, victimhood and risk.

Interface walls were constructed around those areas within which the most excessive patterns of violence occurred. Extensive walling aimed to restrict mobility by sundering the built environment. Without doubt, such structures had the effect of foundation myths, menace and spatial foreboding. As noted by Feldman: '... the wall itself becomes the malevolent face of the people who live on the other side' (1991: 37). Within interface areas senses of hazard and defencelessness were tied to the mobilisation of resistance and all-embracing patterns of cultural cohesion (Feldman, 1991). An essential aim among those who championed ethno-sectarian separation was to subvert customary patterns of intercommunity accord and cohesion (Brewer, 1992; Campbell, 1989). The collapse of the Northern Ireland Labour Party, which drew voters from across the sectarian divide, was tied to the growth in ethno-sectarian animosity at that time. In effect, the growth and the operationalisation of violence were both a direct and indirect assault upon pluralist discourses. Forms of intercommunity loyalty were semantically replaced by the promotion of the argument that the ethno-sectarian 'other' should be viewed as a permanent threat (Shirlow and Monaghan, 2004).

Fear was predicted not only upon the understanding of violence but through the use of urban space to de-territorialise intercommunity connection. For those who acknowledged the need for ethno-sectarian separation, the conviction grew that the ethno-sectarian 'other' was dedicated to harming them. Thus, violent reciprocation was based upon acts that sought to optimise the losses of the 'other' group. For many the impact of violence was to create suspicion and fears of harm, which revealed themselves as ongoing anxiety (Shirlow, 2001). The dynamic of fearfulness was associated with perceptions of threat and the twin forces of resentment and incomprehension (Downey, 2000).

The embedding of ethno-sectarianism was achieved by the construction of discursive formations that created a systematic and conceptual framework capable of defining 'truth'. A significant number of those people living in Belfast have been influenced by processes of ethno-sectarian enclaving and the championing of the 'home' enclave as morally enhanced through the amalgamation of key symbols and distinct discursive practices. Such processes of spatial demarcation echo Foucault's (1979) concern with the field of objects, the subject of knowledge and how 'legitimate' perspectives govern and prescribe the spatialisation of fear and processes of inclusion and exclusion. Ethno-sectarian separation encouraged and re-established discursive formations based upon the enfolding of ideas, the alignment of specific moral inflection and the creation of new mechanisms of power. The perpetual search, by some, for spatial enclosure and socio-spatial demarcation is clearly tied to Sack's notion that the creation of ethno-sectarianised spaces produces: 'Boundaries which are virtually impermeable ... [and which] isolate communities, create fear and hate of others, and push in the directions of inequality and injustice' (1999: 254).

Divisions between illicit and licit discursive moralities conditioned everyday practices via offensive and exclusionary practices. The spatialisation of fear, ethno-sectarian purity and allegiance became understood as both a regulation of community space and a system of ethno-sectarian classification (Jarman, 1997). The evidence, presented in Chapters 4 and 6, on how sectarianised fear and prejudice reduce mobility and intercommunity contact illustrates how social practice still engenders different imaginings of community and the production of community-based and eclectic forms of political identification, fear and violent enactment.

The rise of ethno-sectarian violence in the late 1960s led to more defined physical, visual and political definitions of the boundary between the 'self' and the 'other' community. The separation of populations facilitated the targeting of the adjacent community. Moreover, the presence of the British Army within these boundary zones led to a concentration of violence within such areas. The increasingly strong relationship between residential segregation and the enactment of violence and political discord became contingent upon a series of other relationships such as the environments of everyday life, violence (both imagined and real) and social manipulation (Shirlow and Pain, 2003).

Many people living in interface areas defend their need for physical separation by claiming that it provides them with a degree of security against attack from the 'other' side. Ironically, 'protective' walls were the very places within which frequent, persistent and recurrent, if often low-level, violence occurred. The impact of fear upon mobility has meant that the instruments of violence have subordinated many to the abstraction of ethno-sectarian affiliation.

Map 3.1 above illustrated conflict-related deaths in Belfast and their relationship to interface walls and segregation between 1969 and 2004. Table 3.4 indicates that a third of the victims of politically motivated violence were murdered within 250 metres of an interface, and around 70 per cent of deaths occurred (but representing only 53 per cent of Belfast's population) within 500 metres of all segregated boundaries. In addition, over 80 per cent of deaths occurred within places that were at least 90 per cent Catholic or Protestant.

Table 3.5 shows that 1,417 persons died in the Belfast Urban Area as a result of the contemporary conflict. Of these 922 (65 per cent) were civilians. In relation to combatants, RUC and UDR members made up 5.5 per cent of all deaths compared to British soldiers who constituted 9.8 per cent of all fatal casualties. Loyalist and republican paramilitaries constituted 5.9 and 11.5 per cent of all fatalities, respectively.

Over three-quarters of all civilians (78.3 per cent) were killed in either North or West Belfast. Similarly, 90.6 per cent of British soldiers and 65 per cent of loyalists and 84.7 per cent of republican paramilitaries were killed within the same two geographic areas. East

Belfast contained 9.1 per cent of all fatalities and, as shown in Map 3.1, most of these deaths are located around the Short Strand–Ballymacarrett interface, the only significant ethno-sectarian boundary within that part of the city. The majority of civilians killed in South Belfast were murdered within the city centre. However, pockets of death around the Ormeau Road and Donegall Pass interface are also significant. Evidently, the more prosperous parts of both East and South Belfast endured the lowest levels of politically motivated violence.

Table 3.4   Relationship between politically motivated deaths and interfaces

Distance from interface (metres)	% share of all deaths within BUA
Less than 100	13.47
Less than 200	28.89
Less than 300	44.25
Less than 400	57.28
Less than 500	66.53
Less than 600	71.88
Less than 700	75.91
Less than 800	79.39
Less than 900	81.53
Less than 1000	84.25
Over 1000	15.74

Source: Authors' calculations

Table 3.5   Victim by type and location 1968–2005

	North	East	South	West	Total
Civilian	371	72	128	351	922
Local police/army	21	19	17	21	78
British armed services	40	3	10	86	139
Political activists	7	3	2	17	29
Loyalist paramilitaries	27	19	10	29	85
Republican paramilitaries	46	13	12	93	164
Total	512	129	179	597	1417

Source: Authors' calculations

Such a high rate of violence within highly segregated places indicates the link between factors such as residential segregation, interfacing and social class. It is not surprising that violence encouraged political and cultural retrenchment and the physical and cognitive remapping of the city. The reorganisation of space, due to violence, increased separation and reemphasised the fundamentals of ethno-sectarian 'difference'. As indicated in Figure 3.2, the share of persons who lived in Belfast and died in the city is closely correlated to their home address. Nearly one-third of all victims were murdered within their homes or only a matter of metres from their place of residence. In sum, death within one's own community was commonplace and furthered the idea that violence was based upon an assault upon community. Given the proximity of death to residence, it is evident why conflict-related deaths are understood within discourses of group suffering. The proximity of so many deaths to home places forged strong and at times enduring notions of group-based losses and in so doing diluted the capacity to perceive violence as community-based assail. The parochial nature of violent enactment aided the overall process of territorial entrapment.

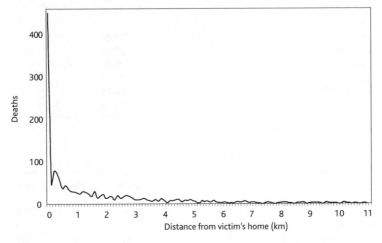

Figure 3.2   Distance between location of conflict-related fatalities and the victim's home in Belfast 1966–2004

Source: Authors' calculations

The initial rise in violence in the 1970s complicated, and provided depth to, what had been noticeable but relatively benign forms of separation. Violence may have become less prominent as the course of conflict altered, but the enactment of the most significant form of intercommunity violence since the formation of the Northern Ireland State polarised ideological conflict and contrasted the experience of communities. Violence sparked the imagination of difference and in so doing created new discourses concerning conflicts of non-violent interest. The general failure to recognise the difference between non-violent and violent conflict has impeded the capacity to build meaningful political progress. The problem was never simply the enactment of violence but that violence gave meaning and legitimacy to wider political, cultural and residential forms of separation. Violence, it should be recognised, was only one form of conflict-influenced behaviour.

THE MEANING OF VIOLENCE

The attempt by politicians and paramilitaries to dominate the ownership of community fears, phobias and traumas has been related to a desire to triumph wider discourses of community cohesion and solidarity. This desire to collectivise the interpretation of harm and to mobilise community consciousness has been a key instrument in the reproduction of conflict. The attempted depersonalisation of death, through the promotion of unidimensional and 'community'-sensed versions of victimhood, remains tied to the overall geography of political resistance and the manipulation of the experience of violence. In determining the impact of violence we present information based upon the focus group and interview work undertaken.

The ability to consolidate identity, however loosely, remains dependent upon the capacity to govern the memory traces of conflict through a series of notions of belonging. It is a process of strategic management that is still based upon convincing sections of each respective community that resistance to the 'other' community is a historical struggle and that the residents of harmed places are the makers of a profound history. This process of identity formation, by means of the selective representation of violence and harm, has been tied to a wider fund of interpretive possibilities.

The importance of controlling the interpretation of the harm caused by violence has become crucial in the definition of intercommunity discord. The recent and significant growth in the erection of community-based memorials testifies to twin strategies of commemoration and the political commodification of landscapes of suffering and endurance. One of the primary reasons for the perpetuation of political discord was the controversy over victims, state collusion and the demand for apologies. The never-ceasing demand for inquiries into deaths and the abuse of human rights is rarely, if indeed ever, based upon a shared intercommunity request.

Conflict created senses of victimhood, which in their most extreme form were based upon discourses of constitutively defined inclusion and exclusion. For some, there remains a less complex and passionate series of relationships between memory and contested legitimacy. As eloquently argued by Aughey:

there existed a pervasive feeling of victimhood, real and imagined, that has been common to the cultures of unionism and nationalism. Emotionally, this condition served to displace responsibility on to others. Politically, it encouraged a helpless attitude of going with the flow, an acceptance that the history of destruction all around was indeed 'natural'. (2005: 11)

Without doubt the emotiveness of monocultural imaginings of victimhood undermines the possibility of progressive political movement. As argued by one respondent from the Tiger's Bay area:

Tell me this. I heard a former IRA man say that less than 50 per cent of their victims were civilians. Do you not think that way of thinking is a bit, sort of, perverted? Do you really believe there were no sectarians in the IRA? Do you really believe that they think our dead are as important as their dead? I don't have to forget what happened.

For a respondent from the predominantly republican Ardoyne community a similar sense of distrust and loathing was also forthcoming. As argued:

Protestants were part of a brutal regiment (sic) aimed to oppress us Catholics. They colluded with the Brits and were only too happy to join the UDR and RUC murder squads. Them ones (sic) groups and the loyalist paramilitaries were the same thing. Some wore uniforms and the others balaclavas, but they were the same people. One day they will have to ask us for an apology for what they did to us.

However, the outplaying of victimhood is relative and can be centred upon various forms of interpretation. Despite the extensive attempts to control localised narratives of victimhood it is evident that the capacity to do so has been blunted by alternative, if concealed, notions of collective loss. As Gramsci (1971) noted, hegemony, in whatever form, is never fully achieved, even within such strongly and passionately defined places. In relation to this supposition there is no doubting that the capacity to render violence as legitimate and unquestioned was never fully endorsed within the communities within which such acts tended to occur.

The deepening of mistrust between communities was not paralleled by a complete conversion of residents within violent arenas towards the unfettered use of force. Evidently, mistrust and violence are interconnected, but they do not necessarily produce a desire for unambiguous violent acts or for revenge. For many who live within some of the most violent places in Northern Ireland, harm is also understood as having placed an unwarranted burden upon both communities. Indeed, a common theme among respondents was that the capacity to adopt narratives of intercommunity dialogue was required as was the need to reject the fiction of blamelessness. There was a strong and defined notion of shared thinking concerning victimhood, but such intercommunity acceptance remains without political, cultural or social recognition. The complexity of the emotions driven by violence and the meaning attached to it were explained by a female respondent from Lenadoon (a republican/nationalist community) in West Belfast:

There were times when people were killed in this community and you felt so much hatred in your heart. But then someone on the other side would die and you would feel ashamed that you had felt that way.

There was an enemy, but not an enemy that you could always hate. You knew your fellow Belfast man or woman was suffering too, just like us. Maybe more than you or maybe less, but you knew they had loss and shame on their side, just like us.

Violence did have the capacity to create a more unified sense of mistrust through the representation of ethno-sectarian assault as an attack upon the 'home' community. When such violence had a destructive impact upon neighbours, family and friends it defined a new relationship within which attacks upon individuals were interpreted as an assault upon the well-being of a whole community. In addition, violence and the fear of being a victim of violence changed the nature of 'everyday life'.

A common theme among those interviewed was that community, prior to the expansion of violence, was viewed through loose connections around faith, family networks and other social norms. Such networks also included intercommunity engagement within which shared interests were played out through camaraderie and friendship, especially with regard to activities such as boxing, pigeon racing, card playing and drama.

What emerged in many interviews among those who lived in interfaced communities was that political identity was more forcefully mobilised once it was recognised as having been violated. It is a form of identity formation that (Durkheim, 1915) recognised as being based upon a general process of ethnic homogenisation and the unfolding of new power relationships. The obvious result of violence was that it deepened the sense of devotion to each respective community and also the pursuit of a community-based sense of protection. As noted by a Protestant respondent on the Suffolk estate:

I never hated the other side. I hate what some of them did to my community. I was never one for the Orange Order or the UDA or anything like that. But when they killed people I knew where my loyalties really were. They was (sic) with my own people. The deaths they caused, them deaths just made me see that I needed to be standing shoulder to shoulder with my own community.

I knew they weren't all to blame, like. But how could I work out which of them was out to get me? It was best to avoid all of them as best I could. I never really really hated them but I could never trust them again.

The complexity of the issues surrounding violence is furthered when it is appreciated that violence produced different emotions among persons living within interfaced communities. For some, as shown above, there was no overall desire to destroy the ethno-sectarian 'other' nor was there a demand to present the sanctity of legitimacy as complete. However, violence is generally interpreted as being based upon targeting, and within a climate of loss the capacity to place the victims of the 'other' community upon the same plane as your own community's dead is rare. As noted by a resident in Short Strand who claimed to be a former member of the PIRA:

The problem is that we always start from whataboutery, you know 'what about this and that'. People died and we still haven't been able to cope with it. For me it has changed. I don't care about the legitimacy of violence any more. Legitimacy and who was right and wrong won't heal this place. Some people still have to have their legitimacy so that they can explain themselves. I fought but the fighting is not now as important as the healing and the need to get on. The healing is much more important, but hardly recognised at all. Surely, the future has gotta be about getting on with a new way of living together.

A series of similar commentaries presented conflict as an experience that was regretted. The strongest and most commonly identified belief was that violence and the fear of becoming a victim encouraged increased forms of ethno-sectarian separation as a means of non-confrontational survival. This is not to deny that many respondents identified violence against their community as being unfounded, unjust and unacceptable. But there was a sense that violence was a form of extreme human behaviour that was set against an incapacity to resolve the conflicts that emerged. Violence was recognised as being controlled through reducing the volume of social interaction between communities. The main and near-universal effect of violence was that it altered the nature of social and spatial interaction. Within a climate of menace and fear, individuals located spatial forms of conflict mitigation. As noted by one respondent from Tiger's Bay: 'The best way to stay safe was to stay out of their area. Even if it meant giving up on aul (old) muckers (friends).'

The decline in violence has not meant, as explored in Chapter 4, that the durability of fear and mistrust has been diluted, as may have been expected within a less harmful environment. The legacy of violence is not merely that it created competing discourses of harm but also that it embedded the logic and need for intercommunity separation. Furthermore, the reproduction of conflict since the beginning of the peace process has been redesigned around more intensified disputes concerning territorial control and ever-present disputes over Orange marches. Such marches are viewed by republicans/nationalists as space-claiming incursions that are centred upon boundary transgressions. For those who support the right to march the ability to utilise 'traditional' routes is sacrosanct. Such dissimilar perspectives undermine the capacity to reposition cultural loyalty within a more pluralist frame.

A glimmer of hope is thus to be located among those who appreciate the dissimilarity between hatred and harm. Despite the monocultural tendency toward commemoration and uneven interpretations over blame it is heartening that few residents within segregated places lionise violent acts. This suggests that spatial behaviour and the desire to remain isolated from the 'other' community are reactive to violence and not merely centred upon a universal sectarian hatred.

However, the fundamental and enduring problem is that violence undermines the capacity to adopt more pluralistic political codes. Violence, and the memory and perpetual experience of it, reduces the capacity for intercommunity harmony and perpetually converts the electorate away from alternative political discourses. Separation is thus based upon the development of resistant frameworks and coping strategies developed through a mix of experience and political engineering. As shown in Chapter 4, not all residents of segregated places are converted to the same practices and beliefs, but for most there remain practical and experiential realities that reduce the desire, of most, to cross the ethno-sectarian Rubicon.

4

Between Segregated Communities

The ongoing violence that has been a feature of political discord in Northern Ireland since the paramilitary ceasefires of 1994 has upheld the maintenance of boundaries between competing political perspectives and segregated spaces. There are several forms and causes of violence experienced within and between the segregated communities of Belfast. Parading has been a catalyst for rioting, especially with regard to the formation of residents' groups who have contested the right to march in places within which the act of parading is held to be offensive.

Events surrounding Drumcree and the Holy Cross dispute triggered an extensive series of riots and other episodic violence performed within the interface zones between republican/nationalist and loyalist/unionist territory. In general, there has also been an ongoing series of violent acts based upon the intimidation of persons living on the edge of segregated communities or those who constitute the minority within such places. Such spatially vulnerable communities have been targeted through stone throwing, paint bombing, hate mail and a systematic pipe-bombing campaign that has been, in the main, orchestrated by loyalist groups (Shirlow, 2003a, 2003b).

There are crucial issues concerning such violence. First, the decline in violence between 1994 and 1998 was followed by a substantial growth in disorder after 1998 owing to the destabilising effect of the fourth Drumcree disputes and the growth in the support base for anti-Agreement unionism. Violence reappointed ethno-sectarian allegiance as a platform of departure and enmity between segregated communities. Second, violence created tensions between communities, especially republicans and the PSNI. The use of force by the PSNI during riots furthered allegations that police reforms had not had a consequential impact upon police behaviour. A central strand in the Belfast Agreement, which aimed to create an agreed policing solution, was undoubtedly undermined by the varied interpretations of what had been upheld as the reutilisation of policing tactics employed by the RUC. It was further evidence that the Belfast Agreement's reformist edge had run into the contentious sands of street-based political confrontation.

An equally important impact of such violence was that it reversed the desire to enter places where the 'other' community was dominant. Interview material suggested that many residents of segregated places, in the period between the paramilitary ceasefires and the fourth Drumcree dispute in 1998, had begun to cross ethno-sectarian boundaries in the pursuit of consumption and leisure practices. Such practices were, it was contended, suspended in the wake of violence, menaces and threats attached to more recent violent acts.

The reappearance of violence furthered, especially among young persons who had been immune to the extensive violence that existed prior to 1994, an ethno-sectarian logic and a strengthened connection between anthropologies of purity and impurity and debates centred on the concept of 'otherness' (Williams and Chrisman, 1993). Violence thus reasserted the need for boundary maintenance and reduced the desire and also the capacity to access communities dominated by the 'other' ethno-sectarian group. The fluid and less sectarian use of space that had been evident prior to violent episodes appears to have been lost as a new series of violent acts 'confirmed' the need for security consciousness. The reappearance of interface violence, as was the case with violence at the beginning of the present conflict, provided the shock necessary to undermine normalised intercommunity engagement and mobility between communities.

Whether imagined or real, it is important that fears and avoidance strategies generated by violence or perceptions of threat are taken seriously because they constitute the manner through which ethno-sectarian decisions and loyalties are fashioned. Avoidance strategies matter because they are real to the people who hold them and as a result of such cognition they undermine the ability to create meaningful political stability (Gold and Reville, 2003; Massumi, 1993; Tulloch and Lupton, 2003).

FEAR, PREJUDICE AND MOBILITY

At one level, the analysis presented below follows a long-standing tradition of ethno-spatial research in which urban segregation is visibly identified by symbolic expressions of group identity and avoidance of an objectified 'other'. The subject matter is frequently understood through analysing the complex relationships between segregated communities that are divided by political identity and ethnic 'choice' (Bew *et al.* 1997; Boal, 1969; Ley, 1994). Violence and the threat of intimidation have created psychological burdens and prejudiced attitudes that are underwritten by restricted mobility between politically antagonistic communities. An authoritative body of work presented by Borooah (1996), Sheehan and Tomlinson (1996), Shuttleworth *et al.* (1996) and McVeigh and Fisher (2002) has indicated that 'chill factors' are attached to the experiences and perceptions of both fear and associated risks.

The recognition that fear and avoidance are influenced by territoriality and ethno-sectarian practice indicates that the abnormal structure of Belfast's consumption, leisure and labour markets can be interpreted through an understanding of the realities of everyday life. For many residents of segregated areas, the ideological role of ethno-sectarian disputation still influences the naturalisation and appearances of social and productive relationships. The perceptions and realities of insecurity that militate against intercommunity mobility are understood as being linked to either a prejudiced position or a perceived or actual fear of entering a community within which an individual or group will be part of an 'unprotected' minority. Evidently, the crossing of ethno-sectarian boundaries is affiliated with wider ethno-sectarian notions within which the 'other' community is viewed as a menacing, repressive and unwelcoming ethno-sectarian construct.

Understandings of threat from intimidation can also be dedicated to behavioural responses that are linked to perceived hazards, dangers and other environmental stimuli. Indeed, many interface zones are festooned with paramilitary regalia and graffiti such as KAT (Kill all Taigs[1]) and KAH (Kill all Huns[2]). Fear and avoidance of the 'other' ethno-sectarian group also connect with wider debates that are attached to the rationality and irrationality of mobility-based decisions. Similarly, the desire to remain within areas dominated by the ethno-sectarian 'self' or in places that are viewed as spatially neutral is based upon the identification and selection of places that are understood as safe, trusted and secure.

The key point is that many people who live within segregated communities endorse and practise the divisive labelling of place because they understand the need to do so. The bounding of space is thus understood within what are comprehended as rational acts and behaviours, even when such decisions are unwarranted and unjustified. Ultimately there is a lack of faith in the 'other' community and the ability of that community to treat outsiders in a positive and constructive manner. Experience of violence, threat and menace can impact upon spatial interaction given that a practical consciousness is influenced by memories, myths, symbolic orders and self-imagery. Each of these cognitive features of spatial choice and other stimuli constitutes an interpretation of safe and unsafe spaces through a practical and at times exaggerated consciousness of situated individuals (Jabri, 1998; Tulloch, 2004).

The nature of such fears and prejudices can also be temporal. It is obvious that the volume of violence increases in the summer months at the onset of the marching period. Youths are also out of school, usually for eight weeks during the summer term, and the violence that they indulge in is set against an obvious ethno-sectarian framework. Episodes and events such as those outlined above can also stimulate emotions such as anger and foreboding, and accordingly the nature of ethno-sectarian enmity may rise and then fall.

MOBILITY AND MARKED SPACE

Our work on ethno-sectarian enclaving and the reproduction of ethno-sectarianised fears within Belfast is based upon two surveys conducted in 2004 on individuals living within interface communities in Belfast. The surveys collected data on households, and in sum information on over 9,000 individuals was forthcoming.[3] The surveys aimed to determine the level of spatial interaction between interfaced Catholic and Protestant communities as well as exploring the nature of intercommunity appreciation and acknowledgement. Given that each of the six pairs of communities share similar socio-economic profiles and are adjacent to each other it would be expected that there would be a similarity between them in terms of public and private sector-based usage and mobility. However, the levels of social, cultural and economic interaction between communities were low, and the reasons for these dissimilar mobility patterns are firmly attached to an emotional landscape of fear, prejudice, loathing, the experience of violence and the more obvious marking of space through hostile imagery and graffiti.

Most respondents felt relatively safe within their own community but had reservations, at best, concerning entering areas dominated by the 'other' community. The lack of interaction between communities was similar and no one group were disadvantaged more than any other. Enclaved communities, places that were surrounded by the 'other' community, suffered most and residents within them generally undertook extensive journeys to access services. Larger and more homogeneous communities, within which an extensive range of facilities is located, produced more internalised mobility patterns.

In general it was found that a mere one in eight people worked in areas dominated by the 'other' community. In addition, 78 per cent of respondents could provide examples of at least three publicly funded facilities that they did not use because they were located on the 'wrong' side of an interface. In some communities, upwards of 75 per cent of the survey respondents refused to use their closest health centre if it was located in a place dominated by the 'other' community.

Over half of all respondents (58 per cent) travel at least twice as far as they have to, usually into neutral areas or areas dominated by co-religionists, in order to locate two or more private sector-based facilities. Eighty-two per cent of respondents whose nearest benefit office was located in the 'other' community's territory used facilities located in areas dominated by their 'own' community even though this meant undertaking significant journeys. These disruptions to everyday living are alarming when it is acknowledged that interface areas contain the most excessive forms of social dislocation and deprivation within Belfast. Somewhat depressingly, the evidence collected also suggested that around one in eight respondents forgo healthcare for younger members of their families rather than use the nearest health facilities when such facilities are located in areas that are perceived to be unsafe.

A mere 18 per cent of respondents undertook, on a weekly basis, consumption-based activities in areas dominated by the 'other' community. It is generally assumed that fear of being attacked by the 'other' community is central in determining low levels of intercommunity contact. However, a subjective reading of such information masks a series of relationships complicated by age, gender and intracommunity threat. Nearly sixty per cent (58.9 per cent) of respondents who did not undertake shopping and other consumption activities in areas dominated by the 'other' community did so owing to what they recognised as fear of either verbal or physical violence. Around 9 per cent stated that they would not undertake such journeys owing to a desire not to spend money in areas dominated by the 'other' community, in what they deemed to be 'loyalty' to their own community. Immobility, as shown in Table 4.1, is a much stronger determinant in the choice of facilities than loyalty to the home community.

Table 4.1   Percentage share of those who do not use facilities in areas located in the other community by age and reason

Age	Fear of the other community	Fear of my own community	Fear of both communities	Loyalty	Other
65+	45.6	20.2	20.2	8.5	5.5
45–64	49.6	16.2	14.2	13.5	6.5
25–44	64.6	9.4	9.2	8.8	8.0
16–24	75.8	8.4	6.2	5.6	4.0
Av.	58.9	13.5	12.4	9.1	6.0

However, 13.5 per cent of respondents would not undertake journeys into areas dominated by the other community owing to fear of being ostracised by their own community. A similar number stated that fear of both the 'other' community and the 'home' community was also a motivating factor in their refusal to enter areas dominated by the 'other' community. Fear of 'my own community' was directly linked to the belief that entering areas dominated by the 'other' sectarian group would lead to individuals being 'punished' by members of their own group. Such fear thus operates as a judgement based both on inter- and intracommunity perceptions. Pensioners were those least likely to perceive the 'other' community as a menacing spatial formation and were thus more likely to cross between segregated communities. For pensioners fear of 'my own community' or both communities was more palpable than it was among those aged 16–44. Relatedly, those aged between 16 and 44 were more likely than their older counterparts to acknowledge fear of the 'other' community as the primary reason for the non-crossing of ethno-sectarian boundaries.

In relation to basic spatial statistics, the survey also aimed to measure the impact of these factors upon consumer and recreational behaviour. Interaction distances between communities were measured as straight lines between the population-weighted centroids of each respective place. The findings regarding the Ardoyne and Upper Ardoyne communities, which are presented below, were reproduced across all of the other communities studied with regard to the burdens created by ethno-sectarian influenced immobility.

With regard to Ardoyne (republican/nationalist) and Upper Ardoyne (unionist/loyalist) the findings obtained reflects the latter's position as a smaller community that lacks retail facilities. Given the close spatial proximity of the communities (0.45 km between the centroids of Ardoyne and Upper Ardoyne), and similar socio-economic profiles, predicted utility maximisation should be very similar in terms of interaction levels. As shown in Map 4.1, Upper Ardoyne contains no conventional shopping facilities and in contrast, less than 0.5 km away, the Ardoyne community contains six large grocers, as well as a range of other outlets. It would be expected that the closest facilities, those located in Ardoyne, would be relied upon. However, only 19 per cent of those surveyed in Upper Ardoyne used these nearby facilities on a weekly basis. Similarly, only 18 per cent of Ardoyne respondents used the nearest leisure centre, which is located in Upper Ardoyne.

A statistical measurement of dissimilarity in consumer choice can be drawn out by a simple coefficient of determination, r^2. Table 4.2 lists highly insignificant relationships for all four service and recreational categories. When comparing mean distances to services, Upper Ardoyne respondents travel twice as far for daily shopping necessities, and over six times farther to pubs and social clubs than their Ardoyne counterparts. The main burden placed upon residents of Ardoyne is the ability to locate safe leisure centre facilities. Most (82 per cent) refuse to use the nearest centre, and instead travel to leisure centres in republican/nationalist areas such as West Belfast. Ardoyne respondents who use leisure centres travel over six times further than their Upper Ardoyne counterparts. The contrast in distance travelled to services is a further indication of the reluctance of the two communities to share local facilities. When distance and volume are taken into account, consumer choice between the two communities becomes even more insignificant, and lower r^2 values are calculated. These are in stark contrast to the idealised r^2 value of 0.982, which represents distances.

Table 4.2   Consumption activities and distance travelled in Ardoyne and Upper Ardoyne

| | Relationship | Total distance of all visits (km) | | Mean distance per visit (km) | | Dist-weighted relationship |
	r^2	A	UA	A	UA	r^2
Daily shopping	0.026	41.5	80.5	0.415	0.805	0.014
Weekly shopping	0.004	141.2	181.8	0.780	0.983	0.043
Pubs/clubs	0.160	43.7	150.9	0.455	2.902	0.014
Leisure centres	0.103	241.1	93.2	10.483	2.824	0.007
Total	**0.001**	**467.5**	**506.4**	**1.169**	**1.369**	**0.002**

Table 4.3 details the distances travelled between paired communities. The Oldpark and Manor Street communities are

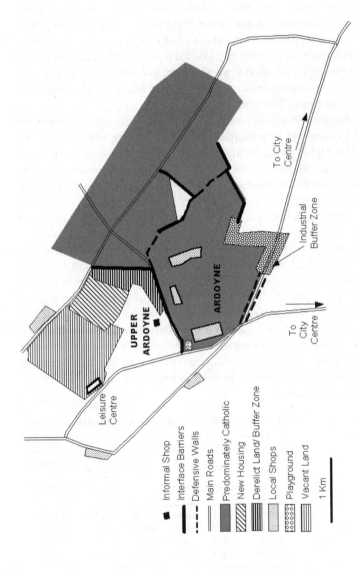

Key:
- Informal Shop
- Interface Barriers
- Defensive Walls
- Main Roads
- Predominately Catholic
- New Housing
- Derelict Land/ Buffer Zone
- Local Shops
- Playground
- Vacant Land

1 Km

UPPER ARDOYNE

ARDOYNE

Leisure Centre

To City Centre

To City Centre

Industrial Buffer Zone

Map 4.1 Segregation, facilities and the two Ardoynes

located between the Cliftonville Road and Crumlin Road in North Belfast. The Oldpark community is predominantly Protestant and is a community in demographic decline. There are few shopping facilities within this community, which is increasingly enclaved by the twin processes of Protestant outmigration and nearby Catholic population growth. Oldpark residents tend to engage in consumption practices in the Greater Shankill area or the city centre. They travel twice as far as their Manor Street counterparts for daily and shopping centre-based consumption. The respondents from Manor Street, a predominantly republican/nationalist community, travel shorter distances for services, with the exception of leisure centres, as they can avail of such services locally in either the Bone district, Ardoyne or Cliftonville Road.

Table 4.3   Consumption activities and distance travelled in Oldpark, Manor Street, Short Strand and Ballymacarrett

	Mean trip (km) Oldpark	Mean trip (km) Manor St	Mean trip (km) Short Strand	Mean trip (km) Ballymacarrett
Daily shopping	2.54	1.30	1.51	1.25
Shopping centres	1.45	0.77	2.54	1.30
Leisure Centres	1.35	1.64	2.79	1.22
Health centres	1.50	0.81	1.23	0.71
Pubs	1.99	1.58	0.65	1.26
Post offices	0.59	0.50	1.10	0.75
Library	1.24	0.71	2.17	1.11
Job centres	1.20	1.24	3.12	1.77
Social security	2.54	1.30	2.98	1.76

The residents of Short Strand, the only republican/nationalist community in East Belfast, endure the same burdens as the respondents from Oldpark and Upper Ardoyne. During the rioting and disorder between Short Strand and Ballymacarrett in 2003, local GPs and pharmacists alleged that they were threatened by loyalists who wanted a cessation of services to residents of Short Strand. In terms of daily shopping the respondents from Short Strand tend to stay within their own areas. Short Strand respondents tend not to use the facilities at places such as Connswater and the Newtownards Road, which are located within a predominantly Protestant place. There is a general tendency for individuals from Short Strand to choose facilities within the city centre or the outer rim of West Belfast. Consequently the distances covered by Short Strand respondents are much higher than those endured by residents from Ballymacarrett. The nearest library, which is located in unionist/loyalist territory, is within half a kilometre from the centre of Short Strand, yet respondents travel on average 2.17 km to access library facilities.

In order to determine the impact of ethnic division between study areas respondents were asked a series of questions that linked labour market issues to understandings of intimidation and workplace choice. As indicated below, it is obvious that fear and prejudice are both produced within the choice of workplaces, and in many instances negotiating 'safe journeys' to places of work is a crucial factor in the search for employment. It is also important to stress that the lack of knowledge concerning workplaces in both neutral places and areas dominated by the 'other' community is also attached to a combination of factors such as car ownership and low levels of mobility more common in lower-income communities. As with many individuals from deprived communities in the UK and Ireland, the knowledge of urban areas is linked to socio-economic well-being. However, the lack of knowledge of places beyond sites of residence is also, unlike similarly deprived communities outside Northern Ireland, influenced by the categorisation of areas dominated by the 'other' community as unsafe and unwelcoming. Evidently, prejudice and a disdain for areas dominated by the 'other' community are not necessarily based upon previous direct experiences of harm or visitation to such places, but that does not mean that ethno-sectarian labelling does not take place. A common feature of doing this type of work is that respondents cannot readily identify communities on maps but they do know the names of places and the ethno-sectarian affiliations of the communities based there.

A wider topic in relation to work and the construction of labour markets is the issue of religious discrimination. In sum, three relationships affected respondents with regard to dissuading active job seeking in areas dominated by the 'other' community. First, it is contended that there 'is no point' in applying to workplaces dominated by employees of the opposite religion owing to active discriminatory practices. Second, journeys to or into areas dominated by the 'other' religion are sometimes deemed to be unwise and potentially dangerous. Third, there is a sense of prejudice and a desire to reduce the potential for contact with the 'other' community.

The extent to which respondents within each study area determined that their community was discriminated against was high. In republican/nationalist areas, 92.1 compared to 78 per cent in unionist/loyalist areas stated that their community is discriminated against when seeking work. Such a high response rate fits into the wider sense that both communities are being treated unfairly. Indeed, much of the labour market debate in Northern Ireland has been fixated on the existence, reproduction or perpetuation of discrimination.

Within the survey the vast majority of respondents in both republican/nationalist and unionist/loyalist communities (81 and 72 per cent respectively) stated that on at least three occasions they had not sought a job in an area dominated by the 'other' community. When respondents of working age were asked to determine if they would work in a place dominated by the 'other' community a complex relationship was established. Around 60 per cent from each community stated that they 'would consider' working in such a place. However, only 15 per cent would do so if the workplace was located in an area dominated by the 'other' group. Thus a location dominated by the 'other' community was a more prominent demotivating factor than working in a place dominated by the 'other' group that was located in a neutral site.

Around 60 per cent of respondents had previously worked in places dominated by the 'other' community. However, only around one in eight presently do so. Moreover, around ten per cent of respondents aged below 35 had worked in a place dominated by the 'other' religion compared to around 74 per cent of pensioners. This decline in the numbers working in places dominated by the opposite religion may signify how the impact of fear, intimidation and 'chill factors' has operated during the contemporary conflict. It could be postulated that people are tending to locate workplaces that are 'safe' more than was the case previously or that they are tending towards more mixed workplaces or, as is most probable, they pursue a combination of both of these factors.

Not surprisingly, ethno-sectarianism plays a dominant role where residents shop and use facilities, and how far they are prepared to travel. The great majority of consumer interaction is between origins and destinations of the same community. The journeys that cross ethno-sectarian boundaries and interfaces were generally ascribed to the use of large and spatially detached shopping centres. Ethno-sectarianism has dominated consumer decision-making and produced a distorted pattern where many residents from one community choose one set of destinations and residents from the other community another, with very little sharing.

It is almost as if there are two distinct sets of consumers, both living within a one km radius, and where the concept of 'competing destinations' synonymous with 'newer' interaction models is virtually confined to distinct ethno-sectarian market boundaries. In classical discrete choice and interaction models, this is akin to allowing fear and avoidance to act not only as friction but also as barriers on distance.

VIOLENCE, COMMUNITY RELATIONS AND CONTACT

Nearly 40 per cent of respondents surveyed stated that they had been victims of physical or verbal violence since the ceasefires of 1994. Overall, 967 incidents of physical violence were recorded, with a near equal split with regard to ethno-sectarian background. Among residents from unionist/loyalist places 82 per cent of these incidents were attributed as sectarian attacks by the 'other community'. Within republican/nationalist places 52 per cent of attacks were denoted as based upon ethno-sectarian assault. In addition, there were ten times as many assaults attributed to confrontation with the police and army within republican/nationalist places. The fact that the nature of violence against each community is dissimilar may well partly account for the higher (93 per cent compared to 71 per cent) levels of perceived residence 'at the interface' within predominately Protestant places compared to republican/nationalist communities. This higher level of cognition regarding the perceived residence at an interface may also be part of wider unionist/loyalist sense of territorial decline and encroachment. Eight out of ten respondents also stated that, since 1994, they had witnessed physical violence by either the 'other community' or the 'security/crown forces'.

With regard to community relations 66 per cent of respondents contacted in 2003 believed that community relations had worsened since 1994. This compares to 52 per cent who held the same position in 1998. There was a related decline from 32.3 to 28.1 per cent, between these periods, of respondents who contended that intercommunity relationships had improved. Respondents from unionist/loyalist places (69 per cent) compared with those from republican/nationalist communities (61 per cent) upheld the proposition that community relations had worsened.

Table 4.4 comprises information on various forms of violence and antisocial behaviour since 1994. The data combines information on each set of communities, but it is worthy of note that respondents from unionist/loyalist communities tended to indicate that all forms of violence and antisocial behaviour had increased. Some of the findings are peculiar given that it is evident that drug use has increased, as has the general level of youth-based delinquency. In recent years, for example, there has been a growth in stone-throwing assaults upon the emergency services and public transport vehicles. When respondents were questioned on why they believed that drug taking and violence within their own community by residents from that community had decreased several interviewees noted that they had not wished to provide any evidence that would suggest that their communities were being affected by internalised forms of antisocial behaviour. This was a common explanation and would suggest that many respondents would not criticise the communities within which they lived in the desire to present a sense of community renewal and cohesion. A common theme also within republican/nationalist communities was that drug use was essentially a phenomenon located within loyalist/unionist places.

Table 4.4   Perceptions of changes in volume of violence and antisocial behaviour since 1994

Type of violence/antisocial activity	Increased	Decreased	Same
Crime	40.8%	29.0%	40.2%
Drug use	30.0%	31.6%	38.4%
Violence within area by own community	20.1%	48.1%	31.1%
Violence within area by other community	63.6%	22.7%	13.7%
Violence within area by youth from own community	28.1%	34.1%	37.8%
Violence within area by youth from other community	58.6%	17.9%	13.5%
Violence against area by other community	73.6%	13.1%	13.9%
Police/Army	52.5%	13.0%	34.5%

However, the same level of caution and consideration was not evident when arguing that there had been significant increases in violence from the 'other' community and the police and army. In the three instances where violence from persons from the 'other' community was measured, the percentage share that acknowledged an increase tended towards significant majorities of respondents. It is obvious in small communities such as White City and Whitewell that there has been a growth in interface violence. However, in some communities where such attacks have declined, such as Lenadoon and Suffolk, the share of respondents who acknowledged an increase was significant. In some communities, such as Ardoyne, Upper Ardoyne, Short Strand and Ballymacarrett, it is clear that interface violence has increased. However, the important issue is that in most of these places antisocial behaviour among fellow residents has probably also increased but without the same level of recognition. It is further evidence that the spatialised form of loyalty to the community-based 'self' can be reproduced within wider attitudes and opinions that aim to select blame through an ethno-sectarian logic.

Most work on fear and mobility has indicated that women tend to be more fearful, particularly after dark, than are their male counterparts. As indicated with this survey 31 per cent of women, compared to 9.8 per cent of men, stated that they felt either scared or would not walk though their community when dark (Table 4.5). However, as shown in Table 4.6, the dissimilarity in the responses received between men and women narrows significantly during the summer months. When questioned on fear during the 'marching season', the number of men who will not walk or who are apprehensive about walking through their local area rises to over 60 per cent, which can be ascribed in particular to a rise in those who feel unsafe or scared. Among women, the share located within the three categories of fear rises to over 80 per cent.

Table 4.5   How safe do you feel when walking through your local area when dark?

	Male	Female
Safe	42.4%	20.4%
Quite safe	20.7%	21.6%
Unsafe	25.2%	27%
Scared	7.1%	21.5%
Wouldn't go	2.6%	9.5%

Table 4.6   How safe do you feel when walking through your local area during the marching season?

	Male	Female
Safe	20.8%	8.2%
Quite safe	8.3%	11.2%
Unsafe	41.7%	19.9%
Scared	24.7%	42.6%
Wouldn't go	4.6%	18.2%

A further merger in the levels of fear between women and men was apparent when respondents were questioned on whether they would 'feel safe when walking through an area dominated by the opposite religion' both during the day and at night. A mere 17.3 per cent of males and 10 per cent of females would undertake such journeys during the day (Table 4.7). In addition, nearly half the males and females simply would not undertake such a journey during the day. The percentage share that would not undertake such a journey when dark rose to 92.2 and 94.2 per cent of men and women respectively. When respondents were asked if they would travel through 'an area dominated by the opposite religion after dark' those who would do so and feel either safe or quite safe were in a distinct minority (men, 6.9 per cent; women, 3.8 per cent).

Table 4.7   How safe do you feel when walking through an area dominated by the opposite religion during the day?

	Male	Female
Safe	9%	3.7%
Quite safe	6.3%	6.3%
Unsafe	23.5%	15.7%
Scared	12.5%	22.5%
Wouldn't go	48.7%	51.8%

When the measurement of fear was broken down and analysed in relation to each community it became evident that there were similarities between the republican/nationalist and unionist/loyalist study areas. In sum, the majority of respondents in each community feel safe when walking through their own community during the day. However, more people in unionist/loyalist areas (52 per cent compared to 43 per cent in republican/nationalist districts) feel either unsafe, scared or are not prepared to walk through their local area when dark. Within both communities, around three-quarters of respondents stated that they feared being attacked by members of the 'other' community. In republican/nationalist areas around 22 per cent cited threat of attack by the police or army as a primary concern. Only 4 per cent in these areas felt primarily threatened by members of their own community. Within unionist/loyalist areas over 20 per cent sensed that attacks by people from within their 'own' community was more problematic than attacks by members of the opposite community.

THE MEANING OF SPATIAL DISCONNECTION

Virtually all of the respondents of pensionable age who were able-bodied used facilities in areas dominated by the 'other' community. Interviews among the pensioner group disclosed that the majority were not afraid to enter 'alien' territory for four main reasons. First, social relationships that existed prior to the contemporary conflict had tended to endure and older people visited the other community in order to maintain social networks. Second, pensioners were three times as likely as the other age groups to have either Catholic or Protestant relatives. This could mean that intermarriages were more common prior to the present conflict and that such relationships produce some form of intercommunity contact. Third, older people tended to be repulsed by violence, which they contended had destroyed a previous society within which community relations were relatively 'normalised'.

Although pensioners conceded that their communities had been victimised by sectarian violence it was also contended that their community had also been involved in transgressive sectarian behaviour. A fourth issue that emerged was that of religious conviction and a belief that it was immoral to judge whole communities as abnormal and unwelcoming. It was apparent, from the interviews conducted with pensioners that alternative social histories, within which there has been an extensive form of intercommunity linkage, diluted the rationale of sectarian sentiment and as a result fear of this prejudice against the 'other' community was tempered by more experienced forms of dialogue and mutual respect.

Stronger and more sectarian attitudes were located among those aged between 16 and 40. Few within this group undertook, by choice, any form of intercommunity linkage or visits to areas dominated by the 'other' ethno-sectarian group. For this group, the experience of residential segregation was channelled through a framework of exclusive and sectarian representations and ideological 'tradition'. Sectarianism and fear of the 'other' community were not viewed as a repressive relationship, but as an articulatory process that enshrined the need for spatial segregation. Space, for those who articulate sectarian discourses, was seen to function as an object that hosts historical 'truths' and collectivised forms of communication.

Community and history thus served as micro-territorial constructions, which reinforced the manner through which the presentation of hostility was a valid and necessary sectarianisation of space. Among those who advanced unwavering sectarian attitudes, the actuality of residential segregation via spatial constructs and sectarian behaviour was imperative in order to operate and further topographic conflicts. During conversations with those who upheld sectarian narratives it was acknowledged that social space is definitively coded through a sectarian discourse. It was commonly understood that urban space is purposely assembled by both ideological and physical separation. However, one of the most pronounced factors that distanced the sectarian group from those who were non-sectarian or less sectarian was the manner in which they eulogised, through the expression of devotion, the communities within which they lived. Each member of the sectarian group imagined their community in terms of notions of utopian integrity, loyalty, kinship and purity. In comparison, non-sectarians were more likely to accept that 'their' community was more complex and that they contained multiple forms of division and transgression.

The sectarian group also voiced impassioned sectarian narratives and a defined and pronounced sense that the 'other' community was abnormal, antagonistic and uncompromising. Overtly sectarian arguments asserted that their community had been victimised and that the perpetuators of such attacks were odious representations of the 'other' community. The refusal to identify that analogous violence was a burden upon the 'other' community meant that ethno-sectarian separation was comprehended through an unwavering and unidimensional ethno-sectarian logic.

Fear, within this group, was explained through the framework of reproaching the contrary community. As a result, sectarianised readings were constantly linked to an acknowledgement of spaces of fear and the location of unsafe places. Within such a climate of ethno-sectarian cognition and telling, intercommunity contact was discussed through tales of violence, aggression and the 'other' community's violent transgressions. Violence from within the 'home' community, which had been directed at the 'contrary' community, was articulated as a strategy of defence or as a rational form of reprisal.

Paradigms of ethno-sectarian purity and the location of impurity within the 'other' community are predicated upon social relations to such an extent that vocabulary necessary to challenge ethno-sectarian practice is itself undermined. This implies that telling, violence and the reproduction of fear are based upon sectarianised relationships that aim not only to reproduce residential segregation but also to undermine belief systems that identify ethno-sectarian purity as a socially constructed and imagined set of politicised relationships. Maintaining the ability to manage the propaganda of ethno-sectarian belonging is very often undertaken through reminding fellow residents that the 'other' community is to remain mistrusted, thus further ensuring that sectarianised relationships will continue through the endorsement of imagined morality and sectarianism. Space, for those who articulate sectarian discourses, is seen to function as an object that hosts historical 'truths' and collective discourses. Community and history, for this group, serve as micro-territorial constructions, which reinforce the way in which geography presents sectarian hostility as a valid politicisation of space. As stated by a male respondent aged 42 from Short Strand:

I've had terrible beatings from Prods [Protestants]. A lot a sum the lot of them snout [Protestants] bastards. They hide behine the peelers [police] and their sectarian murdering thugs. You see them, they have no politics nor nothing to offer anyone. Just a hatred of us. Not on bit ah difference whether there rich or poor, they're all the same shower ah sectarian fuckers.

Similarly, for a male resident aged 48 and from Upper Ardoyne, the following was concluded:

I would love to burn those bastards out down there (Ardoyne). The soap dodgers [derogatory name for Catholics] breed like rabbits. All you ever hear from them is whinge, whinge, whinge. Why don't they get jobs and live like decent people?

It was also evident that few interviewees who articulated sectarian attitudes had fashioned any significant or long-lasting intercommunity-based relationships. More alarmingly, respondents who endorsed a sectarian consciousness viewed involvement in intercommunity schemes as perfidious and as a betrayal of community-based loyalty. Some even argued that spending money in places dominated by the 'other' group was duplicitous. In certain cases, buying goods in the 'other' community was directly linked to paramilitaries and violence against the 'home' territory. As stated by a resident from Upper Ardoyne:

If I knew that my neighbour was shopping in Fenianville [Ardoyne] I would take a pounder [a hammer] and knock his head of his shoulders. Those shops down there give money to the RA (PIRA).

Similarly, as stated by a respondent in Ardoyne:

One of my neighbours bought a suite of furniture from a place in the Shankill [Protestant area]. I told him that I wouldn't be in his house as long as that furniture was there. Like, he was giving money to people who [had] attacked us!

Around one in five interviewees undertook visits on a daily basis to areas within Belfast dominated by the 'other' ethno-sectarian group. Members within this group generally felt unthreatened by such visitations but would at time suspend intercommunity mobility during periods of tension. This group, most of whom articulated anti-sectarian arguments, shared a similar profile of violent abuse from the 'other' community to those located within the larger sectarian group. However, they were twice as likely as their sectarian counterparts to be victims of verbal threats by members of their 'own' community. Many stated when interviewed that such menacing threats had occurred owing to respondents articulating either anti-sectarian discourses or 'insolently' challenging the activities of paramilitary groups. The capacity of this group to undermine the power of community belonging and in so doing challenge the orthodoxy of ethno-sectarianism, through delivering new narratives of intercommunity camaraderie, has ensured that they remain among sectarians as a distrusted and disliked out-group. This is unfortunate in that the non-sectarian group possess the ability to deliver an alternative process of 'telling' within which the sectarian construction of space is identified as being part of a process of control and ethno-sectarian practice.

The non-sectarian group were nearly three times more likely than their sectarian counterparts to work in places dominated by the 'other' ethno-sectarian group. This finding tended to be supported by a sentiment that individuals had some form of obligation to challenge intercommunity divisions. All held meaningful friendships with members of the 'other' ethno-segregation. Reasons stated for being less sectarian were eclectic. Some said their anti-sectarianism was based upon religious conviction, and others that an interest in particular types of music or social activities had drawn them into positive relationships with members of the 'other' ethno-sectarian group.

Some suggested that sexual relationships had altered sectarian attitudes. For others, self-education and an interest in left-wing ideology were pinpointed as catalysts in the challenge of sectarian animosities. However, non-sectarian respondents were not apolitical. Virtually all interviewees from republican/nationalist communities voted for Sinn Fein, but in each instance insisted that it was noted that they were opposed to violence. Similarly, unionist/loyalist respondents all supported unionist/loyalist politics, but as with their republican/nationalist counterparts were against the use of violence. However, the factor that was most commonly mentioned among the non-sectarian group was an acknowledgement that they could no longer discuss their intercommunity contacts or anti-sectarian discourses publicly. Each stated that they had been berated when neighbours and family members had found out that they had consorted with members of the 'other' ethno-sectarian group.

In many interviews with the non-sectarian group it was acknowledged that respondents did not want 'others' in their 'own' community to know that they engaged in a range of intercommunity activities. Most respondents stated that they would rather lie to their neighbours than tell them that they had been socialising with members of the 'other' ethno-sectarian group. In terms of politics, the majority of these respondents were critical of the sectarianism within the political parties that they supported but would not discuss this with people they did not trust. Some stated that they would repackage the goods they bought in shops that were obviously purchased in the territory of the 'other' religious group. As explained by a respondent from Upper Ardoyne:

We shop in Curlies [located in Republican West Belfast]. It's so cheap there and who is going to know we are Prods [Protestants] if we go there? But we take Tesco bags with us and put the shopping in them before we go home. We make sure that we dump the bags from Curlies in case someone sees them in our bin.

If I walked up that path with Curlies bags I would get my windies [windows] in [broken].

In combining the attitude of pensioners and those who are non-sectarian it was evident that a sizeable minority of respondents held dissimilar attitudes with regard to overt and sectarianised interpretation of affiliation and habituation. However, given the hostility that is directed towards those who are non-sectarian by members of their own community, it is evident that few arenas exist within which to actively and openly articulate non-sectarian beliefs.

5
Coasting in the *Other* City

At the beginning of the conflict the poet John Hewitt challenged the political complacency of a middle class disconnected both politically and spatially from the urban nadir of sectarian strife. He shrewdly cast a sarcastic eye upon a disinterested middle class in the following extract from his poem 'The Coasters':

> You coasted along
> to larger houses, gadgets, more machines,
> to golf and weekend bungalows,
> caravans when the children were small,
> the Mediterranean, later, with the wife ...
>
> The cloud of infection hangs over the city,
> a quick change of wind and it
> might spill over the leafy suburbs,
> you coasted along too long.

The middle classes have generally negotiated their way through conflict, and if anything the children and grandchildren of the 'coasters' have been socially bolstered by the rise of the contemporary service economy and an expanded consumption–production nexus. As the evidence presented in this chapter suggests, social and economic restructuring has not produced a new political class, even one motivated by self-interest and material pursuits. While they are less atavistic and more malleable than their working-class counterpart, they have not constructed a shared identity capable of challenging the binary politics of Northern Ireland.

The period since the first paramilitary ceasefire in 1994 has been characterised by strong economic growth accompanied by distinctive social and spatial unevenness with regard to labour market outcomes (Russell, 2004). There are new patterns of consumption and an attempted reimaging of city life in an effort by the British state to uphold middle-class lifestyles and to present Northern Ireland as a normalising place. Even though this process of 'normalisation' has been unfolding for several years, there have been few analyses of what local reforms and global economic, social and technological shifts have meant with regard to the transformation of interests and identities (Guelke, 2005). While the processes of change have been understood by economists and regional analysts, there has been less commentary on the way in which these shifts have impacted upon social structures, demography, identity and lifestyles (Neill, 2004).

The debate on housing in Northern Ireland has been interpreted within the origins, reproduction and resolution of grievances that tested the very legitimacy of the state. The empirical focus upon mapping segregation, researching its effects and exploring the responses of housing agencies has dominated the literature. Other social processes around increasingly mixed tenure patterns have been ignored or their significance denied. Yet the preoccupation with social exclusion and the emergence of an underclass has dominated European, national and even local policy initiatives on urban regeneration. Even here, they were connected to the state's unrestrained but unsatisfied attempts to buy off nationalist/republican working-class discontent (Murtagh, 2002).

Mixed housing places were identified as a priority within the Belfast Agreement, a tangible sign of stability and what Aughey referred to as part of the 'narratives of progress' in the post-ceasefire period (2005: 161). Mixed communities are empirically significant as their capacity to resist the inexorable rise of segregation over time and place and the preparedness of some, outwardly at least, to live at peace with one another is of importance. While not extensive, the stock of mixed housing straddles both tenures and regions across Northern Ireland. Certainly, they are dominant in less urbanised markets and are peopled by the higher social classes, but their creation, durability and potential provides an important counterbalance to the segregated communities looked at elsewhere in this book.

Questions remain about the meaning of residential integration in highly racialised urban settings. Ash Amin (2002) made the important point that it is not inevitable that ethnic groups 'mix' in areas where the statistics suggest high levels of integration. Given the focus of this book, it is important to unpack the nature of socio-spatial mixing in the city and to assess the extent to which alternatives to segregation are real and hold potential beyond a small number of privileged neighbourhoods. This chapter is concerned with the concept of de-ethnicisation with regard to middle-class identity formation and whether it finds spatial expression in Hewitt's 'leafy suburbs'. The empirical focus is the suburban expansion of Carryduff and Outer South-East Belfast (COSE), which consists of the wards of Galwally, Beechill, Carryduff East and West, Hillfoot, Cairnshill, Knockbracken, Newtownbreda and Wynchurch. In 2001 it had a population of 28,263 and is broadly divided between the established, largely middle-class community that developed in the 1950s and 1960s and new greenfield developments that began in the mid-1980s.

'NEW' IDENTITY FORMATIONS

The relaxed regime for property capital in the 1980s was only part of the explanation for the development of outer south-east Belfast. Just as production pressures resulted in new land for development, consumption pressures ensured a steady demand for new dwellings. The privatisation of social consumption, especially with regard to housing, speeded up the trend towards suburbanisation and the growth in social mobility. These processes led Elliott to detect a:

fracturing of old monolithic Catholic/Protestant identity, a fracturing, which particularly when the Protestant community, created on a deep sense of decline and despondency, was the backdrop to the escalation of Protestant paramilitarism before the two cease-fires. That fracturing, however, may be the necessary precondition for recognition of what unites rather than what divides the communities. (Elliott, 2002: 181)

The general contention of Elliot's work was that middle-class Catholics would be less convinced about the objectives of Irish nationalism and would see material benefits of membership in a benign, if uncertain, Northern Ireland. For their part, the unionist middle class would, it was supposed, be less dependent on British state guarantees. Both, it was assumed, were prepared to accept Northern Ireland as a place apart and agree territorial coexistence with unwritten rules and procedures in places such as south-east Belfast. Obvious symbols of identity were not encouraged by a community driven more by the collective protection and enhancement of property prices and less by atavistic ethnic claims.

While sectarian animosity is still visible among all social classes, a growing body of evidence supports the thesis that the middle classes, irrespective of their religious affiliations, increasingly share similar lifestyles and socio-economic pursuits, which are mutually agreeable and inherently less antagonistic (Shirlow, 1997). The argument has been advanced that shared class values and lifestyles are becoming an increasingly important variable politically as well as socially and that they cut across ethno-religious formulations of national identity. As shown below, there are a range of mixed environments to be located within middle-class space, but this does not mean that ethno-sectarian identities are not reproduced or that ultimate allegiance does not remain wedded to unionism or nationalism.

One of the most significant changes in recent times has been the growth in a Catholic middle class. Douglas identifies this group as having evolved via new forms of labour market growth and a series of packages and initiatives designed to produce mutual understanding and ethnic closure through the past three decades. These included the establishment of the Fair Employment Commission, anti-discriminatory legislation, the promotion of integrated education, new community relations policies and administrative reform. For Douglas, local and global processes helped to create a 'new place world in Northern Ireland' (1997: 166). As he expressively argued: 'New social structures have influenced the process of socialisation and the greater diversity of social behaviour has led to more subtle definitions of the Self and the Other' (1997: 171–2).

Other analysts of class modification have argued that the British state has aimed to manage middle-class lifestyles and politics in an attempt to solve wider conflictual tendencies within Northern Irish society. Since the imposition of Direct Rule the state has developed administrative, legal and policy initiatives to maintain its overall legitimacy:

This aim has been pursued through a policy of socio-political normalisation and the adoption of practices whose primary goal is to secure the construction of a set of social relationships which, it is hoped, will transcend sectarian hostilities and engender socio-economic normality. (Shirlow, 1997: 99)

Shirlow (1997) argued that British state's management of the conflict extended to building new class alliances, so, while it saw limited scope in working-class urban areas, it deliberately and consciously attempted to bind the middle class to dependence on government resources, political guarantees and, most crucially of all, publicly funded jobs. The expansion of the service and public sectors coupled with rigid policies on fair employment helped to grow a significant Catholic middle class (Osborne and Shuttleworth, 2004).

Fair employment legislation, the encouragement of employment growth in the service sector and the increasing importance attached to the equity agenda have helped manipulate political affiliations in an attempt 'to forge a third tradition, capable of living with evident cultural and political ambiguities' (Shirlow, 1997: 100). The shift in labour market demand towards the supply of a better-educated workforce has also enabled those with a strong educational base to fare better. Furthermore, it is evident that social mobility among Catholics has removed them from the experience of labour market discrimination. Using time series data on labour mobility, Miller concluded that:

Unlike the past generations in 1973/74 data, if Catholics and non-Catholics begin their working lives with the same levels of education and first job, their mobility through their careers will not be directly advantaged or disadvantaged by religion. (2004: 63)

Nearly one-third of the jobs created in Northern Ireland between 1990 and 1995 were professional, managerial or administrative positions. The Catholic share of new employment grew by 32.8 per cent between the mid-1970s and mid-1990s, and over 70 per cent of the successful applicants for new positions had tertiary-level education and came from the middle classes (Shirlow, 1997).

Class mobility created an increasingly divided and discordant material set of experiences for middle- and low-income Catholics. More affluent Catholics have tentatively embraced a degree of rapprochement toward the British state (Graham and Shirlow, 1998). This rapprochement is a consequence of rising social mobility in which education, fair employment and increased public sector jobs helped to forge 'a new Catholic middle-class' (Breen and Divine, 1999: 56) distinctive from the 'old' Catholic middle class, which was essentially located within health, the licensed trade, education and legal services.

O'Connor (1993) argued that Catholic tertiary-level education provided the instrument through which this community rose through the class ranks. In the early 1970s, 70 per cent of Queen's University students were Protestants, but recently more than 50 per cent of the student body is Catholic. The trend for young Protestants to emigrate to English, Scottish and Welsh universities has been accompanied by significant numbers settling in Great Britain. It is a process of settlement that testifies to the Protestant middle class feeling more British and one that guarantees that the Catholic middle-class position within the labour market will continue to grow. The location of new mechanisms of power, for this newly mobile Catholic community has also been identified by O'Connor:

The speed with which Catholic lawyers, doctors, accountants, and entrepreneurs of various kinds have developed access to political decision-making and their way into an economic mainstream, once closed to them, has left nerves jangling inside the Catholic community and beyond it. (1993: 16)

The Protestant middle class have also been affected by increased social mobility. In cultural terms they have tended to remove themselves from the Orange Order and Masonic lodges as linkage with such organisations no longer upholds the same status within a professional community detached from traditional industries. Professional and middle-class unionists are now able to pursue clear objectives through the alternative arena of civic society and via participative rather than representative institutions. Professional bodies that lend voices to the concerns of the province's bourgeoisie no longer require endorsement from the electorate of Northern Ireland in order to negotiate with the Northern Ireland Office. Indeed, these pressure groups are often more likely to have the ear of government ministers than the representatives of political parties, which often enjoy substantial democratic mandates (Coulter, 1999).

Middle-class unionists are now able to pursue their interests outside narrow sectoral politics in a way that is ostensibly non-aligned. But their concern for political stability is class-related and aimed at achieving the conditions of renewed accumulation in Northern Ireland. This has also involved building closer relationships with the economy of the Republic of Ireland.

An increasingly important facet of middle-class life is non-voting. With regard to the Agreement, in particular, it was found that the middle classes, especially unionists, voted to push though the Belfast Agreement, but only a month later many failed to turn out to vote for the pro-Agreement UUP and SDLP in the same manner. Non-voting, in particular, recognises certain realities for the middle classes. First, they feel removed from the issues of marching, state and paramilitary violence and are not spatially aware of the discord that remains. Second, they have located material affluence and cannot identify reasons why voting is important to them. Third, some see the growth in political polarisation as obviating the ability to create a strong 'third' party tradition.

As shown below, although most middle-class people may 'feel ambivalent about parts of their allegiance, they feel bound to choose' (Morrow, 1996: 58). For Boyle and Hadden (1994), labelling is also tied to subtleties of identity. With regard to these subtleties they identify three main groups:

- those who have no direct association with either the main community, notably those from Britain and abroad, or who have deliberately abandoned any communal association;
- those who have close associations and sympathies with both communities, notably those involved in mixed marriages or with other close relations in both communities;
- the larger number of people who can readily be allocated to one or other community for some purposes but who reject many of the prevailing attitudes in those communities and who want to be left to get on with living peacefully together.

However, Breen and Divine conclude that there remain significant differences:

The notion that the two communities in Northern Ireland mirror each other is quite false. Members of the Catholic community display a considerable degree of heterogeneity in attitudes to community relations issues, political preferences and constitutional preferences whereas members of the Protestant community tend to share a greater commonality of outlook in these areas. (Breen and Divine, 1999: 63)

There are also allegations that the movement of Catholics into middle-class-dominated communities has the effect of creating some angst within certain established Protestant communities. It has been contended that Protestants have moved out of certain areas owing to what they identity as a 'greening' process or the spatial creeping of Catholics into 'their' communities. O'Connor quotes a Protestant's reading of the arrival of middle-class Catholics in his area:

The Protestant middle class feels threatened – not by the IRA but by the sight of Catholics arriving – so what do they do? They send their children to England and then follow them, or they head for North Down. What do we get? A middle-class ghetto in south Belfast, demanding bigger schools, bigger churches, next a Gaelic football field I expect. (quoted in O'Connor, 1993: 191)

MIXING AND HOUSING

Gallagher (1994) showed how the Catholic middle class was relatively dispersed compared to Protestant counterparts within the city. A quarter of the middle-class Catholics living in Belfast live in the north and west of the city close to major concentrations of working-class Catholics. However, Gallagher also found that there was no spatial correlation between working- and middle-class Protestants in the city. He also found that the majority of middle-class Catholics lived in mixed areas while the majority of middle-class Protestants lived in segregated areas.

The expanding middle-class search for housing opportunities has been dedicated to a narrow band of wards largely to the south of the city (McPeake, 1998). McPeake found that Catholics search for houses in much smaller spatial markets and for longer periods of time than Protestants. Catholic housing search is therefore confined to fewer compatible suburban properties compared to their Protestant counterparts. Wards with a rising Catholic representation include Malone, one of the most affluent parts of the city. In 1971 one in eight residents in Malone were Catholics compared to one in three in 1991 (Cormack and Osborne, 1994: 82). A symbol of this shift has been the upgrading of St Brigid's Catholic Church on the Malone Road. The church is known locally as the Maid's Chapel, with many members of previous congregations being servants within affluent Protestant homes. Today a new and much larger church stands near the site of the original church and possesses one of the most affluent congregations in Ireland.

Cormack and Osborne (1994) also identified another important growth thrust in Castlereagh where the proportions of working-class Catholics had declined particularly between 1981 and 1991 but where middle-class Catholic representation did increase, particularly in new housing developments to the outer part of the borough. They pointed out that Beechill (17.4 per cent), Knockbracken (32.8 per cent), Four Winds (44.0 per cent) and Newtownbreda (22.8 per cent) are all areas of Catholic middle-class expansion. Work by Lloyd et al. (2004) has summarised the stubbornness of segregation and picked out significant areas of ethno-religious mixing. The 2001 Census showed that of the 171 wards in the Belfast Metropolitan Area, 20 (12 per cent) had a population more than 70 per cent Catholic, 112 (65 per cent) were more than 70 per cent Protestant and only 39 (or 23 per cent) were between these polls.

NICE AND COSE

The analysis presented here on middle-class mixing is concerned with the housing search processes and decision-making patterns of people in Carryduff and Outer South-East Belfast (COSE). In evaluating the issue of identity other factors, such as housing search patterns and housing values are analysed with regard to the formation of the contemporary study area.

It is contended that both Protestants and Catholics share some behavioural and attitudinal processes, housing histories and social profiles. The research was based on a household survey of 1,000 household heads in the study area and a matching survey of 1,163 people in Northern Ireland to allow benchmark comparisons to be made. For instance, an analysis of the age profile for Catholics and Protestants with the rest of the Northern Ireland population shows some distinctive features with regard to COSE residents.

Typically, heads of household are more likely to be found in the middle age ranges reflecting the connection between housing mobility and life cycle. However, two-thirds of Catholic household heads in COSE were aged 25–44 compared to a Northern Ireland average of 39.0 per cent. COSE has been directly affected by new forms of social mobility among younger Catholics. Protestant household heads tend to be older and are most likely to be found in the 45–64 age group. In COSE a total of 45.5 per cent of Protestant household heads were in this age group compared with 32.2 per cent for the Northern Ireland sample as a whole. This suggests that Protestants moved into the area much earlier than their Catholic counterparts, a phenomenon that backed up the contention that widespread social mobility for Catholics came more towards the end of the twentieth century.

Social class differences are also pronounced. A total of 27.1 per cent of Catholic household heads in Northern Ireland are now in the top two social class groupings compared with 55.9 per cent of the COSE population. The figures for Protestant household heads are 27.3 and 44.7 per cent respectively. In Northern Ireland as a whole 26.5 per cent of Catholic heads of household are in the bottom two social class groupings compared with just 18 per cent of Catholics for the COSE population. Again, the comparative figures for Protestants are 27.8 and 32.5 per cent respectively. Catholics (68.4 per cent) within the COSE study were more likely than Protestants (48.3 per cent) to have lived in mixed areas before moving into the study arena. Protestants were also more likely to have lived in segregated areas (45.5 per cent) than Catholics (18.3 per cent).

Proportionately more Catholics (37.3 per cent) than Protestants (27.4 per cent) were interested in buying in the right area, whereas more Protestants tended to favour the right house (11.5 per cent), more so than Catholics (6.0 per cent). As Figure 5.1 shows, Catholics were more likely to search in mixed religion areas (56.1 per cent) than Protestants (20.6 per cent). This correlates with McPeake's (1998) findings, which showed that Catholics engaged in longer and more discriminating approaches to house purchase than Protestants because of the comparatively fewer housing options in the Belfast urban area. Both religions were keen to avoid segregated housing alternatives.

Both Catholic and Protestant respondents were aware of the religious composition of COSE before they moved there. A total of 58.7 per cent of Protestants had a 'certain' or 'good' idea of the religious composition of the neighbourhood and, emphasising the previous point about housing search patterns, 71 per cent of Catholics knew that COSE was mixed. High and rising house prices, coupled with new build and resale opportunities were equally important. Significant numbers of Catholics (91.2 per cent) and Protestants (88.1 per cent) thought that the price of their house would rise higher than inflation. The 'tradition' of investing in a growth-oriented and value-based housing market is shared equally by both groups, the strongest 'tradition' shared by both communities.

There were also distinctive features about the composition of the labour market in COSE. A quarter of Protestants and 42.2 per cent of Catholics were educated to degree standard, and both Protestants and Catholics were more likely to be in full-time work than was the case in the wider Northern Ireland cohort. In COSE, 61.7 per cent of Protestants and 78.4 per cent of Catholics were in full-time work compared with 54 and 55.2 per cent for the rest of the Northern Ireland sample. The picture that emerges is one of a young, economically active labour market, especially among Catholics. There is also evidence that Catholics have benefited from the expansion of public sector jobs in particular. Catholics are over-represented in nationalised companies and other public sector areas while Protestants tend to be more likely to work in the private sector. When public sector employment within COSE was assembled, 49.5 per cent of Protestants work in this sector compared with 62.7 per cent of Catholics.

The analysis below uses multivariate statistics to interpret an attitude scale in order to draw out some underlying strands in social attitudes and to compare these between the study area and the Northern Ireland population as a whole. A 13-point scale was used based on work by Lewis (1999), itself a reflection of the three elements of identity by Poole (1997). Poole suggested that measures of identity

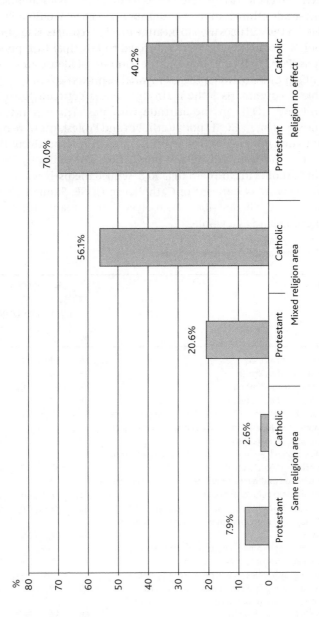

Figure 5.1　Area of housing search

should capture data on ethnicity, religion and nationalism. What was of interest to Poole was the shift between ethnic and civic nationalism implying a de-ethnicisation over time or between generations. Table 5.1 sets out the variables used to measure the three corners as suggested in Poole's model. Results were subjected to Principal Component Analysis (PCA) to try and map out the contours of identity and how they might differ between the two spatial contexts.

Table 5.1 summarises the main differences between Northern Ireland and COSE on the attitude variables. Three points are significant. First, in COSE, both Catholics and Protestants were more likely to describe themselves as British and Irish respectively than the rest of the Northern Ireland population. For example, 29 per cent of Catholics in Northern Ireland described themselves as British compared with 44 per cent of Catholics in COSE. Similarly, fewer Catholics in COSE described themselves as Irish than Catholics in Northern Ireland as a whole.

Table 5.1   Comparison in agreement variables

Variable	Agreement			
	COSE		Northern Ireland	
	P	C	P	C
	%	%	%	%
I think of myself as Irish	39	84	31	90
I think of myself as British	95	44	91	29
I think of myself as Catholic	2	97	2	96
I think of myself as Protestant	96	2	91	1
Political solutions should be through peaceful process only	96	92	91	1
Protestants should learn more about historical events that are especially important to Catholics	83	81	77	80
Catholics should learn more about historical events that are especially important to Protestants	85	82	80	82
I believe in God as the Heavenly Father who watches over me and to whom I am responsible	85	87	86	91
I try hard to live my life according to my religious beliefs	82	82	85	91
I would not mind if my child married someone from the other religion	64	87	56	79
Getting income tax down is more a priority for me than the political future of Northern Ireland	16	15	21	20
The housing authorities should do more to promote integrated housing for Protestants and Catholics	73	80	73	82
Integrated education should be the only option for children in Northern Ireland	69	64	55	53

Second, for Protestants, British ethnicity was stronger in COSE than the rest of Northern Ireland. This would seem to corroborate evidence that the Protestant middle class sees cultural assimilation to a mainstream and marked sense of identity as a strategy for coping with the political uncertainties of Northern Ireland (Coulter, 1999).

Third, both Protestants and Catholics in COSE feel that each should learn more about the 'others' history compared to the Northern Ireland population more generally. Catholics had more positive attitudes to intermarriage than Protestants, but supportive views are held more strongly in COSE than for the Northern Ireland sample. Interestingly, Protestants in COSE are more in favour of integrated education than Catholics but both religions support integration more strongly than in the rest of Northern Ireland.

There is little difference between the spatial and religious differences when integrated housing is considered. It has been noted before that Protestant reading of integrated housing differs from that of Catholics (Murtagh, 2002). This is especially the case in places such as COSE where a growing, wealthy, young and upwardly mobile Catholic population contrasts with an ageing and more indigenous Protestant community. The third point relates to the importance and consistency attached to religion as part of identity. Belief in God, self-analysis as 'Protestant' or 'Catholic' and living life by a sense of religious code were identified by both groups.

Further evidence of mixing was located in terms of marriage and schooling. A total of 13.1 per cent of households in COSE were mixed-religion compared to a Northern Ireland figure of 4.1 per cent (NIHE, 1998, Table A4.13). A total of 7.7 per cent of residents attended a mixed school, and this was evenly distributed among both Protestants and Catholics. Moreover, 15.2 per cent of households had a child who attended an integrated school in Northern Ireland.

Around 14 per cent of Catholics and Protestants had children attending a mixed school. Moreover, 39.1 per cent of Protestants and 57.3 per cent of Catholics said that their circle of friends was comprised evenly of members of both religions. Similarly, 41.1 per cent of Protestants and 39.6 per cent of Catholics believed that their children's friends were mixed-religion in equal proportions. However, as with Northern Ireland, more generally, it is evident that the vast majority of marriages and a significant share of school choices are predicated upon the maintenance of ethno-sectarian boundaries. These latter and enduring trends indicate a significant gap between sharing space and more fulsome and important modes of integration.

Table 5.2 indicates that four principal components were extracted and that they had similar factor loadings. The component matrix is shown in Table 5.3. The first factor reflects a broadly Protestant and pro-British profile, which for the purposes of convenience has been termed *Orange*. On both variables, and between scales, this had high correlation coefficients with variables 1 and 3. The second factor scores highly on Catholic and Irish variables and has been termed *Green*. Factor 3 emphasises the desire of Catholics and Protestants to learn more about each other's culture and history. Factor 4 emphasises a more practical approach to integration by promoting mixed housing and education.

Table 5.2   Variance explained

Component	Description	Northern Ireland Variance (%)	COSE Variance (%)
1	Orange	27.409	21.010
2	Green	20.478	17.644
3	Practical	12.544	11.972
4	Cultural	9.387	9.133
Total variance explained	–	69.818	59.759

Table 5.3   Component matrix for Northern Ireland and COSE

Variable	Component Northern Ireland				Component COSE			
	1	2	3	4	1	2	3	4
1	.748	−.367	.020	.020	.628	.020	−.190	.030
2	−.497	.704	.020	.020	−.436	.020	.549	.170
3	.736	−.493	.245	.020	.694	.020	−.548	.050
4	−.565	.714	.020	.020	−.617	.20	.639	.020
5	.493	.485	.020	.20	.328	.030	.314	.159
6	.643	.442	−.288	−.436	.556	−.173	.638	−.438
7	.579	.479	−.256	−.473	.525	−.169	.683	−.484
8	.319	.430	.725	.202	.338	.747	.289	.163
9	.317	.409	.750	−.030	.279	.776	.265	.177
10	.545	.020	−.370	.237	.455	−.258	.030	.369
11	.020	.181	.020	.646	−.107	−.193	.030	−.020
12	.532	.364	−.190	.337	.380	−.279	.278	.400
13	.347	.320	−.381	.455	.194	−.372	.333	.551

What is interesting about COSE is that it has much lower factor loadings for both the Orange and Green factors when compared with the Northern Ireland sample, although the scores are similar for cultural and practical factors (Table 5.3). The factors explain 69.82 per cent of the variation in the Northern Ireland data compared with 59.8 per cent in COSE. In other words, there is more 'bleeding' of identity in COSE. We see positive and interesting features of intercommunity acknowledgment in the COSE sample, but also signs that differences both predominate and are important.

ACTIVITY AND THE REPRODUCTION OF PLACE

The analysis also measured the volume and pattern of social interaction in the study area. It looks in turn at cultural, sporting, church, social and work-related activity and the frequency with which each sphere of interaction was conducted on a weekly basis. Table 5.4 indicates that Protestants were more likely than Catholics to engage in local cultural activities, and that this was especially the case with regard to the Women's Institute, musical activity and bridge. The mean activity for Protestants (4.86) was nearly twice that for Catholics (2.86) against the activities measured.

Table 5.4   Cultural activity by religion

Activity	Protestant		Catholic		Total	
Cultural	N	%	N	%	N	%
Bridge	6	1.3	6	1.8	12	1.3
Conservation	1	0.2	2	0.6	4	0.4
Drama	2	0.4	5	1.2	7	0.4
Local history	3	0.4	2	0.7	5	0.4
Musical	9	2	3	1	12	0.4
Language	0	0	1	0.3	1	0.1
Women's Institute	13	2.9	1	0	14	1.4
Total	34		20		55	
Mean	4.86	1.03	2.86	0.80	7.8	0.6

With regard to sport (Table 5.5) there were obvious differences. No Protestants participated in GAA activities, in line with obvious cultural and political linkages and the incapacity to learn about such sports without a Catholic schooling. Protestants were much more likely to participate in lawn bowling, which was historically tied to Protestant churches and traditional industries. Soccer was a more common and shared sporting activity, as was keep fit and golf. However, these may be taken separately from each other.

Table 5.5 Social capital and activity by head of household and partner: sports

Activity Sports	Protestant		Catholic		Total	
	N	%	N	%	N	%
Bowling	42	9.4	11	3.3	57	6.1
Golf	61	13.8	49	15	131	14.1
Soccer	23	5.3	23	7.1	51	5.5
GAA	0	0	37	11.3	39	4.2
Tennis	6	1.3	3	1	12	1.3
Watersport	3	1.3	8	2.6	18	1.9
Dancing	0	0	4	1.3	6	0.6
Keep fit	55	12.2	36	10.9	101	10.6
Total	197		180		436	
Mean	17.91	4.06	16.36	5.03	39.64	4.23

Not all activities included

Again, as shown in Table 5.6 there were important differences by religion, with rates of Protestant activity for the Mothers' Union (5.8), the choir (6.2) and Sunday School (5.3) especially high. Catholics were exclusively involved in the St Vincent de Paul (1.6) but comparatively even numbers within both religions were engaged in church cleaning, finance and lay ministry.

Table 5.6 Social capital and activity by head of household and partner: religion

Activity Church	Protestant		Catholic		Total	
	N	%	N	%	N	%
Mothers' Union	26	5.8	1	0.3	29	3.1
Choir	28	6.2	6	1.8	36	3.8
Sunday School	24	5.3	0	0	27	2.8
St Vincent De Paul	0	0	5	1.6	5	0.5
Inter church	13	2.9	3	0.9	18	1.9
Church cleaning	9	2.0	8	2.6	17	1.8
Finance	14	3.1	8	2.4	22	2.3
Lay minister	6	1.4	6	1.8	13	1.4
PTA	12	2.6	12	3.9	24	2.6
Scouts/Guides	12	2.6	10	3.9	22	2.2
Total	144		59		213	
Mean	17.9	3.19	3.0	1.92	21.3	2.24

Social drinking was conducted on a comparatively frequent basis by both religions although Catholics were more likely to go drinking outside the area and in a mixed context. Table 5.7 indicates that 77 per cent of Catholics compared with 54 per cent of Protestants go out for a drink. This may well reflect the dissimilar age structures and more relaxed attitudes towards alcohol consumption within the younger cohort. Going for a meal was a more common experience for both groups as was the attendance of a play/concert.

Table 5.7   Social capital and activity by head of household and partner: social

Activity	Protestant		Catholic		Total	
Social	N	%	N	%	N	%
Drink	252	54	269	77	651	65
Meal	434	92	329	94	924	93
Cinema	210	45	210	60	526	53
Play/concert	255	54	204	58	552	55
Visit friend	423	90	90	95	918	92
Visit relatives	437	93	93	97	941	94
Total	2011		1685		4512	
Mean	335	71.3	280.8	80.1	752	75.3

Activity in the professional sector saw remarkably similar profiles for Protestants and Catholics. A total of 126 people engaged in professional-related activities including 57 Protestants and 53 Catholics. Table 5.8 accounts for a virtually equal profile when trade organisations, unions and professional institute membership were examined, although the numbers involved are very small.

Table 5.8   Social capital and activity by head of household and partner: professional

Activity	Protestant		Catholic		Total	
Professional	N	%	N	%	N	%
Trade organisation	3	0.6	3	0.9	8	0.8
Trade union	21	4.8	17	5	46	4.8
Professional	33	7.4	33	9.9	72	7.2
Total	57		53		126	
Mean	19	4.27	17.6	5.27	42	4.27

THE RELIGIOUS CONTEXT WITHIN WHICH ACTIVITY TAKES PLACE

The research also measured whether the activity was carried out with members of the same religion or in a mixed setting (Table 5.9). The data showed that entertainment activity was most likely to be conducted in a mixed setting (mean = 288.33) while an average of 213.67 people conducted their activity in a Protestant setting and an average of 109.67 people sought entertainment in a mainly Catholic environment. The percentage difference shows that 38 per cent of entertainment activity was mixed but 28 per cent was carried out in a mainly Protestant setting and 15 per cent in a mainly Catholic context. Sports activity had quite high rates of mixed usage (281, mean = 25.55). As noted, Protestant segregated usage was high for lawn bowling (20) while exclusive use of the GAA (39) was recorded for Catholics. However, there were high rates of mixing within golf (105), keep fit (75) and watersports (18). Again, the percentage figures show the extent of mixing in sporting activity. Nearly two-thirds (65 per cent) of activity occurred in a mixed setting compared, with 14 per cent in an exclusively Protestant context and 13 per cent in an exclusively Catholic context.

Table 5.9   Variations between activity and destination

Activity		Protestant		Catholic	
		Mean	%	Mean	%
Cultural	Activity	4.86	100	2.86	100
	Own religion	2.29	47	0.43	15
	Out religion	2.57	53	2.93	85
Sports	Activity	17.91	100	16.3	100
	Own religion	5.45	30	5.09	31
	Out religion	12.51	70	11.2	69
Church	Activity	14.40	100	5.9	100
	Own religion	14.30	99.3	4.7	80
	Out religion	0.10	0.7	1.2	20
Social	Activity	335.1	100	280.8	100
	Own religion	213.6	64	109.6	39
	Out religion	121.5	36	171.1	61
Professional	Activity	19.0	100	17.67	100
	Own religion	1.67	9	0.33	2
	Out religion	17.33	91	17.34	98

High rates of mixing also occurred around professional activities. Very little activity was exclusively Protestant (mean = 1.67) or Catholic (mean = 0.33) and a total of 84 per cent of all activity was in a mixed setting. Protestant segregated usage was higher in the sphere of socio-cultural activity (mean = 2.29) than for Catholics (mean = 0.43). Looked at in percentage terms, 29 per cent of usage was in a Protestant setting, 5 per cent in a Catholic setting and 53 per cent in a mixed environment. Clearly, church-related activities created higher rates of segregated use. This was especially the case where activity data had established different profiles between religions. For instance, the Mothers' Union, Sunday School and to a lesser extent, the choir, had mainly a Protestant usage profile. In general, the use of church infrastructure was higher for Protestants (14.30) than for Catholics (4.70). If usage was considered by percentage it shows that only 9 per cent of all church-related activity happened in a mixed environment. Actual differences in own religion and out religion usage are also shown in Table 5.9. Out religion activity is strong for both Protestants and Catholics in the spheres of professional and trade links, sports and, to a lesser extent, entertainment. Clearly in-group activity is highest for both religions when church-related activity is considered although self-containment is also a feature of socio-cultural activities among Protestants.

CHILDREN AND SOCIAL CAPITAL

A similar approach is taken to the analysis of social capital and children. Table 5.10 shows the activity profile for the eldest child under 16 years of age. Three domains of activity were identified, including socio-cultural, sports and church/school-related. Mean activity was higher for Catholic children in the first two spheres but considerably lower when church- and school-related activities were considered. The church-related sphere was the dominant area of activity for both Catholics and Protestants, with 110 activities registered, 46 per cent of the total. However, most of these activities were conducted by Protestant children (71 per cent) compared with Catholic children (25 per cent). The usage patterns were reversed when sports were considered. Sporting activity accounted for 106 or 44 per cent of the volume measured, but of this total Catholics accounted for two-thirds (69, 66 per cent) and Protestants for one-tenth (24, 10 per cent) of activity. Participation in soccer and the GAA accounted for most of the difference here. Socio-cultural usage patterns, by comparison, were lower, with only 11 per cent of the total and relatively equal numbers of Catholics and Protestants engaged in these activities.

Destination data are shown in Table 5.11, which underscores the extensive segregated use of church-related activities. Scouts and Guides were mainly conducted in a Protestant setting (40), although there was some usage of Catholic (12) as well as mixed (13) settings. Sunday School activity was centred entirely on Protestant-only settings. In terms of sports, soccer activity was sourced in an entirely mixed context, although GAA was exclusively Catholic. Dancing was carried out mainly in a Catholic setting, reflecting the development of a significant infrastructure around various aspects of Irish identity. Finally, drama activity was also centred on mainly Catholic facilities and services. This reflects a broad dichotomy in the assembly of children's social capital that promotes cultural and sporting aspects of Catholic identity and religious-linked aspects of Protestant identity.

Table 5.10   Social capital and activity by eldest child

Activity	Protestant		Catholic		Total	
Cultural	N	%	N	%	N	%
Total	12	8.6	13	7.6	26	7
Mean	3	2.15	3.25	1.9	6.5	1.75
Sports						
Soccer	7	5.1	18	10.5	30	8
GAA	1	0.7	19	11.1	21	5.6
Total	24	17.3	69	40.4	106	28.4
Mean	2.4	1.73	6.9	4.04	10.6	2.84
Church						
Scouts/Guides	42	30.4	19	11.1	65	17.4
Sunday School	33	23.9	2	1.2	38	10.2
Total	**78**	**56.4**	**25**	**14.6**	**110**	**29.5**
Mean	**19.5**	**14.1**	**6.25**	**3.65**	**27.5**	**7.38**

Selected activities only

Table 5.11   Social capital by religion of destination for eldest child

Activity	Protestant	Catholic	Mixed	Total
Cultural				
Total	5	10	10	26
Mean	1.25	2.5	2.5	6.5
Sports				
Soccer	3	6	21	30
GAA	0	21	0	21
Total	**9**	**46**	**48**	**106**
Mean	0.9	4.6	4.8	10.6
Church				
Scouts/Guides	40	12	13	65
Sunday School	38	0	0	38
Total	**80**	**16**	**14**	**110**

The correlation between Protestant sports activity and a Protestant destination was .836 while the equivalent coefficient for Catholics was .792. There was a strong correlation between Catholic use and a mixed destination for the socio-cultural sphere (.934), but caution needs to be exercised as the figures are relatively small.

The reproduction of social relations around church and school domains is highlighted in Table 5.12, which correlates activity with destination data. Protestant church-related activities had a correlation of .990 with Protestant-based facilities and Catholic activity had a correlation of .980 for a Catholic destination of that use. However, there was strong correlation between Catholics and use of mixed-religion facilities (.967).

Table 5.12 Correlation analysis between children's activity and destination

Activity		C Destination	P Destination	M Destination
Cultural	P	.419	.926	.713
	C	1.00**	.117	.934
Sports	P	.144	.836**	.644*
	C	.792**	.288	.426
Church/	P	.558	.990**	.681
School	C	.980*	.603	.967*

*   Correlation is significant at the 0.01 level (2-tailed)
**   Correlation is significant at the 0.05 level (2-tailed)

Table 5.13 shows that there was comparatively little 'leakage' for Protestant children when church/school activities are considered, although there were high rates of own religion use for Catholic sports and socio-cultural services. This again reflects the nature of what are deemed to be 'British' sports such as cricket, hockey and rugby, and what are considered to be 'Irish' sports such as gaelic football, camogie and hurling. In more recent times there has been a dramatic growth in Gaelic games in response to the growth in nationalist/republican confidence and the appearance of extensive television coverage.

Undoubtedly soccer in middle-class areas and other sports such as netball and basketball are intracommunity in nature, although this does not mean that they are played together. As a result 63 per cent of Protestant activity was conducted outside the group compared with just 23 per cent for Catholics. Protestant children were also more likely to source facilities outside their group for socio-cultural activities (58 per cent) than Catholic children (23 per cent). Evidently, many Catholic children are indulging in Gaelic games at the expense of intracommunity-based sports. This situation is unlikely to alter given the nature of the sports played in schools and their religious composition.

Table 5.13   Variations between children's activity and destination

Activity		Protestant		Catholic	
		Mean	%	Mean	%
Cultural	Activity	3	100	3.25	100
	Own religion	1.25	42	2.5	77
	Out religion	1.75	58	0.75	23
Sports	Activity	2.4	100	6.9	100
	Own religion	0.9	37	4.6	67
	Out religion	1.5	63	2.3	33
Church/School	Activity	19.5	100	6.25	100
	Own religion	20.0	103	3.50	56
	Out religion	−0.5	−3	2.75	44

The main concern of this chapter has been to explore a set of distinctive socio-spatial processes around the suburbanisation of outer south-east Belfast. New private developments were fuelled by an emerging upwardly mobile class that had benefited from the expansion of the service sector, public sector jobs and anti-discriminatory legislation. Yet the prospect that social actors in the province might prefer to construct themselves in terms of gender, sexuality, age or more individual inclination is blithely ignored in politics and identity analysis. The most important distinctions that exist in Northern Ireland's growing middle-class society – namely those that pertain to class – are simply written out of existence (Coulter, 1999).

The data show that home owners in COSE differ from the rest of the Northern Ireland population in several important respects. They enjoy advantageous positions in the labour market, have gained material wealth through home ownership and rapid house price increases, and their identities are less uniform and predictable than those of other people in the province. Home ownership, and the benefits accrued from it, is vital to understanding this complexity, and the shared ground is comprised of those who can balance their cultural identity with citizenship entitlements and needs.

It would be wrong to assume that this new socio-spatial class has forged an alternative cultural and, importantly, political identity. While there is tolerance of the 'other', not least to anchor the material advantages of suburban life, these interests are not expressed as a political class or in a coherent set of ideas that make them distinctive in the political and identity landscape of Northern Ireland. Disconnection from the less agreeable realities of segregation and poverty and a capacity to negotiate around harmful political, social and economic territory make them different from but not a radical alternative to the spatial sectarianism of contemporary Belfast.

6
Workspaces, Segregation and Mixing

Danny McColgan, a Catholic postal worker, was shot dead in 2002 by the Ulster Freedom Fighters at his place of work in a predominantly unionist/loyalist area. His death and the related graffiti that accompanied his murder were a deadly threat to Catholics to stay out of unionist/loyalist-controlled arenas. The undoubted aim of such acts and a whole series of other violent enactments has been to stimulate fear and discourage the sharing of employment within segregated places (Shirlow, 2003b).

The desire of many within Belfast to cross ethno-sectarian boundaries was and continues to be undermined by the impact of ethno-sectarian violence and custom (Osborne, 2003). The violence that emerged in the late 1960s had the effect of further segmenting an already polarised labour and housing market, both through identifiable patterns of harassment and the subsequent production of manifest fears and prejudiced avoidance tactics. These fears and prejudices were largely based upon a mode of reasoning that predicted that locating oneself inside an area dominated by the 'other' ethno-sectarian group was injudicious.

However, there is no exact measurement of who is affected by a segregated labour market in terms of residence and social class. At best, there has been an acknowledgement that employment-based segregation impacts most upon those from segregated areas despite 'evidence' of greater mixing within the wider labour market. As with many issues that are affected by segregation there has been a desire to present information, such as greater labour market mixing by religion, that indicates forms of social and economic 'normalisation'. There is no doubt that there are significant signs of increased workplace mixing, but the extent to which this is related to policy and changing attitudes is unknown.

De-industrialisation, for example, has hollowed out highly segregated labour market sectors, but the impact of this is relatively uncertain. It is also evident that other factors such as qualifications and labour market knowledge also impact upon the capacity to gain work as well as the interpretation of safety consciousness.

Somewhat peculiarly, there is a body of work that acknowledges that segregation is still an important variable but there has been little desire to examine the impact of chill factors in a comprehensive manner. It is also acknowledged that the data employed to measure workplace mixing lacks robustness and disguises the level of internal workplace segregation. Yet again the shortcomings exposed in official monitoring data have been inadequately evaluated.

In policy terms, given the failure to adequately explore labour market segregation and chill factors, it is clear that spatial policies fall short of fully appreciating the social meaning, effects and costs of segregation. In examining these issues, it is obvious that there is a need to develop policy instruments and indicators in order to determine the capacity to develop neutral spaces and encourage not only workplace mixing, but forms of workplace mixing that stimulate a greater mix of employees by residence type, religion and standard occupational classification.

As shown in Chapter 4, ethnicised space, with regard to employment, is widely understood as being influenced by the location of lived and working environments and the complex material practices that link the two. As indicated below, residential segregation and workplace location still provide a powerful medium for the articulation and solidifying of communal sentiment and the choice or ability to choose places within which to work. As noted by Green *et al.*: 'The safest conclusion seems to be that for many workers things have got better but that for some people, mainly residents in the inner city, fear remains a significant factor' (2005: 319).

The location of employment in 'neutral' or 'safe' environments has the capacity to challenge certain aspects of ethno-sectarian convention (Freeman and Hamilton, 2004). This does not mean that workplace mixing has the capacity to undermine ethno-sectarian allegiance but that the provision of sites of employment that encourage mixing stimulates the capacity for greater labour market equality and access. Hopefully, the sharing of workplaces provides the capacity to address wider allegations concerning religious/political discrimination (Fraser, 1995; Moody, 2001).

This chapter also aims to determine if there are more effective ways of providing meaning to the collection of data upon workforce segregation. At present religious monitoring data provided by the Equality Commission of Northern Ireland (ECNI) merely provides evidence on the religious composition of companies, but not the composition at each of the sites that they may operate. Without the type of analysis presented here it is impossible to determine, for example, the spatial link between community background and the occupational classification of employees in order to impart a more sophisticated understanding of the depth and reality of workplace segregation.

Within this chapter, measurement techniques are combined with statistical analysis so as to allow for the width of numerous factors such as residential segregation, distance travelled to work and occupational classification. The analysis presented revolves around the principle of measuring distance as a factor within the analysis of residential and workplace segregation and mixing via a unified statistical measurement. Our concluding argument is that sectarianised places continue to undermine the capacity for social justice, equity and political progress. In making such an argument we begin by explaining how contested policies, adopted in order to remove discrimination, and the relationship between community and social exclusion, have not clearly identified how the location of employment reproduces underlying ethno-sectarian structures and practices.

'DISCRIMINATION' AND THE DRIVE TO 'EQUALITY'

Evidently, space matters in terms of understanding the 'choice' of workplaces in environments within which ethnic and/or racial difference is reflected in the labour market. Extreme links between ethnicity and workplace construction are tied to wider discourses and violences that evoke ethno-sectarian and discriminatory practices that seek to maintain the labour market dominance of a particular group (Agnew, 1993; Cohen, 1985; Mouw, 2002; Tewdwr-Jones, 2003). Within violent ethno-sectarian societies it is clear that force, suspicion and mistrust still influence the choice of work among many, especially those on low incomes who live in the more residentially polarised communities (Barnett *et al.*, 2001; Hodgson, 1988; Parker, 2001; Reskin *et al.*, 1999; Sorensen, 2003).

The relationship between religion and political opinion, dedicated to examining employment and unemployment within Northern Ireland, is generally based upon the measurement of the unemployment differential between Catholics and Protestants. Monitoring recent figures from the Equality Commission shows that the steady increase in Catholic participation of approximately 0.5 per cent per annum that occurred throughout the 1990s has levelled off, and for a third year in succession there was no increase in Catholic participation within the workforce. Catholic representation in the private sector stands at around 40 per cent even though Catholics make up 43 per cent of those available for employment. In addition, the same data set shows that within firms with more than 25 employees there has been a decline in Catholic participation.

Recent Labour Force Survey reports indicate that the unemployment rate for Catholics (8.3 per cent) was substantially higher than that for Protestants (4.3 per cent). Among females, 6.1 per cent of Catholics were unemployed compared with 3.9 per cent of Protestants. Within the male group, 9.9 per cent of Catholics compared with 4.7 per cent of Protestants were unemployed. The overall 'unemployment differential' (i.e. the ratio of the unemployment rate of Catholics compared with Protestants) was 1.9 for both sexes combined. Between 1991 and 2005 and despite the introduction of 'tougher' fair employment legislation, and initiatives such as 'old' and 'new' Targeting Social Need, the unemployment differential for Catholic males dropped by a mere 0.2 per cent. For nationalist and republicans such data indicates that community relations and equality strategies have been generally unimpressive.

Unionists tend to make several key arguments concerning allegations of discrimination. First, they argue that accusations of discrimination, made by Sinn Fein in particular, are both imagined and exaggerated. It is contended that there is 'no reason to doubt the overall fairness of the labour market' (Nesbitt, 2005). Unionists agree that the unemployment differential has not narrowed significantly and that this proves that 'rigorous' anti-discrimination laws are ineffective. The inability of anti-discrimination practices to narrow these differentials is seen as 'proof positive' that the reason for these disparities lies elsewhere (Adair et al., 2000). This is usually couched around arguments that concern Catholic birth rates, low levels of outmigration, benefit traps and low skill levels. The overall argument follows that labour market growth is more effective than anti-discrimination policies in creating more equitable shares of employment. As noted by Dermott Nesbitt, an Ulster Unionist Member of the Legislative Assembly and former junior minister within the Northern Ireland Assembly:

There is no *a priori* link between unemployment differentials and discrimination … In simple terms the Catholic community is being told that it is discriminated against and at the same time the Protestant community is being told that it is a discriminating group. Both communities are incorrect in their perceptions. (cited in McVeigh and Fisher, 2002: 37)

However, more vociferous arguments have been made by the DUP, who claim that the impact of anti-discrimination policies has been to engender an unfair treatment of Protestants. This is based upon the finding, in terms of the total workforce, that between 1992 and 2002 there was a fall of almost 5,000 Protestant employees while the number of Catholic employees increased by 22,000. According to Gregory Campbell MP, such evidence is a symbol of a declining Protestant position. As contended:

The answer provides a devastating blow to those who have believed the myth of discrimination against Roman Catholics, and proves that the action which is required is how to ensure that more Protestants get jobs, because it is they who have been losing out. This can only mean that, as I have argued for many years, new jobs are being allocated disproportionately to Roman Catholics.

No other rational explanation can be given for the Protestant/Roman Catholic 'gap' narrowing by 26,000 jobs in a ten-year period.

Now that the myth has been swept aside, the government must begin addressing the reality of Protestant disadvantage in the jobs stakes. (2002: 1)

Republicans and nationalists counter these arguments, arguing that the employment differential between Catholics and Protestants is based upon the impact of previous and present discriminatory practices. A key argument held by nationalists and republicans is that a significant share of employment remains within workplaces that are dominated by a majority Protestant workforce. Thus, Catholics are more likely to be influenced by issues of personal security and ethno-sectarian fear. They also point to the inability of anti-discrimination legislation to actively pursue employers who operate highly segregated workplaces as an example of 'toothless' anti-discrimination policies. As noted in an editorial in *An Phoblacht*, the prominent Republican newspaper:

There are commitments in both the Good Friday Agreement and Programme for Government to eradicate such inequality. We now need to see concerted action so that this legacy of discrimination can become a thing of the past. Attempts to claim that the differential is unconnected to discriminatory practice and policy need to be challenged by all those responsible for the implementation of the Agreement. Rhetoric and words count for nothing unless we see resources specifically targeted against the sectarian inequality in workplaces. (6 May 2001: 1)

After the collapse of the Stormont regime in 1972 the British government undertook a series of legislative reforms through the creation of the Fair Employment Commission (FEC). This attempted, in the 1976 and 1989 Fair Employment Acts, to weaken the authority of discriminatory practices. Evaluations that critically examined the impact of the fair employment legislation were produced by the Standing Advisory Commission on Human Rights (SACHR) in 1987. This work provided a detailed set of proposals both for legislative change and a series of policy initiatives (McCrudden *et al.*, 2004). Crucially, the report shifted the terms of the debate from a concentration upon the eradication of prejudiced discrimination towards a stronger emphasis upon reducing unjustified structural inequality in the employment market, whether caused by discrimination or not.

The combination of state-led political 'normalisation', the SACHR studies and international pressure arising as a result of the MacBride[1] campaign in the United States led to the passage of the Fair Employment Act of 1989. The Fair Employment Act (NI) (1989) intended to provide a more active approach to the practice of fair employment and was to be backed by strong enforcement bodies. Affirmative action was a key feature of the 1989 Act and involved adopting practices that encouraged fair participation while stopping practices that restricted it. In theory, if an employer decided that affirmative action was necessary, then goals and timetables were set that could be used to measure progress in terms of workplace mixing. There were three major forms of affirmative action covered in the Act, which related to:

- training in order to achieve fair participation;
- the encouragement of job applicants in the interest of fair participation;
- the negotiation of agreed redundancy schemes in order to maintain fair participation.

A significant body of legislation now covers a broad swathe of actions that constitute sectarian harassment as illegal in the context of fair employment legislation, via provisions that make it unlawful to treat people 'less favourably' than others or to cause them 'detriment'. Employers, even if they are unaware of sectarian harassment within their workplace, are thus liable for any unlawful acts of sectarian harassment committed by their employees in the course of their employment. Employers are also responsible for the prohibition of all offensive materials and acts that give offence or cause apprehension among particular groups of employees. Other legislation, such as the Protection from Harassment (NI) Order 1997, explicitly states that causing fear is an offence. Clause 6 (1) of the Protection from Harassment (NI) Order states that:

a person whose course of conduct causes another fear, on at least two occasions, that violence will be used against him shall be guilty of an offence if he knows or ought to know that his course of conduct will cause the other so to fear on each of those occasions.

McVeigh and Fisher argue that the ECNI's own deliberations indicate that the promotion of an anti-harassment agenda did not significantly reduce the appearance of sectarian harassment within workplaces. As noted by McVeigh and Fisher: 'Fair employment intervention ... managed to at least recognise that harassment was a problem ... This was, however, a distance from actually reducing the incidence of sectarian intimidation in the workplace' (2002: 12).

In 1997 SACHR identified that anti-discrimination legislation could only be part, though a necessary part, of the process of government addressing the problem of employment inequality. In relation to government-funded job creation the SACHR report of 1997 argued that the government should acknowledge that the Industrial Development Board[2] ignored the impact of 'chill factors'. Chill factors referred to the avoidance of places of workplaces on fears concerning personal security in terms of accessing such places of work or working in places within which employees and/or employers were hostile to a particular group. SACHR also recommended a public review of 'chill factors' within certain areas. Unfortunately, the issue of 'chill factors' did not feature in future policy-making initiatives. This was unfortunate in that the failure to measure the link between religious/political affiliation and the recognition of workplaces that were deemed to be unsafe undermined a more rational interpretation of how violence impacted upon workplace choice. Moreover, the failure to pursue the link between workplace location and wider patterns of segregation undermined a key factor that encouraged labour market inequality and spatially produced forms of social marginalisation.

POLICY, INEQUALITY AND SPACE

Osborne (1996) has pointed out that the development of both fair employment and social need programmes in a top-down tradition of policy formulation and delivery was within the distinctive centralised architecture of the Direct Rule[3] state. Government responses to increasingly vocal external and internal political pressure led to a much wider debate about the role of the state in reinforcing fairness in related areas of public policy (Ellis, 2001).

Policy Appraisal and Fair Treatment (PAFT) was introduced in 1994 in order to eliminate unlawful discrimination or unjustifiable inequality and actively promote fair treatment through government policy-making and implementation (NIO, 1994). PAFT also extended the scope of equality beyond religion to include a wider range of potentially discriminatory categories, including gender, political opinion, marital status, 'having or not having a dependent', ethnicity, disability, age and sexual orientation.

The establishment of new devolved structures brought PAFT out of the policy domain and framed it as a constitutional issue as well as legally binding regulatory practice. The Northern Ireland Act (1998), which provided the legal interpretation of the Belfast Agreement, set down formidable equality duties on all public bodies in the region. The Act gave added weight to the concept of policy-proofing and *ex ante* appraisals and brought legal force to the status of the nine equality categories under Section 75 of the Act.

In carrying out their functions, public authorities must also 'have regard to the desirability of promoting good relations between persons of different religious belief, political opinion or racial group' (Section 75(2)). Government departments had to prepare Equality Schemes that showed how these objectives would be met through current policies and Equality Impact Assessments (EQIAs) were introduced to proof all new programmes against the needs of the nine groups identified in Section 75(1). Equality Schemes describe the specific functions and policies of the department, the policies to be subjected to EQIAs and a timetable for implementation. The schemes devote considerable attention to monitoring, publicity, consultation and describing complaints procedures. Training and skills enhancement has been aimed at developing a greater understanding of the operation of the scheme rather than a wider appreciation of conflict and the role of policy in reducing inequalities per se. A key component of the Equality Scheme is to determine policies that need to be subjected to a full EQIA using appropriate screening mechanisms.

Like the Equality Schemes themselves, EQIAs have, in some cases, been characterised by a reductionist model designed to comply with the legal minimum rather than developing a proactive stance on the causal relationships that reproduce workplace segregation or other forms of spatialised inequality (Ellis, 2001). Causal relationships such as the impact of residential segregation and violence upon mobility were either ignored or mentioned in a highly rhetorical manner. How social exclusion is spatialised and reproduced via territorial practices was also not actively included in the preparation of equality and social need plans.

There was considerable overlap and indeed confusion about the respective roles of PAFT and TSN. TSN was essentially aimed at bending public sector expenditure towards social disadvantage, in part, to close the economic differential between Catholics and Protestants in the region (Murtagh, 2002). Osborne pointed out that this left the policy open to interpretation and question: 'So is TSN aimed at tackling social need irrespective of its distribution between the two communities or is it the primary aim the reduction of differentials in socio-economic experiences between Catholics and Protestants?' (Osborne, 1996: 190).

As with PAFT, the Belfast Agreement provided the impetus for TSN to be evaluated and reworked. The remodelled initiative, called New Targeting Social Need, had three elements, including tackling unemployment, addressing social need (in health, education and housing) and coordinating the actions of government departments through a new commitment to Promoting Social Inclusion (NTSN Unit, 1999). Some Action Plans made vague commitments to researching the causes and effects of deprivation, while monitoring conditions over time was a recurring theme of most departmental approaches. Some plans made specific commitments on inputs, such as where their activities and resources would be invested, while a few of the better models tried to explain how the outputs of policies would result in some form of social and economic closure. As with the other policies outlined above the link between violence and avoidance between communities and the impact of discrimination upon workplace choice was omitted.

To understand the subtle interweaving of location, territory and social need more interpretative methods are required. As is indicated below, we must understand better the way in which segregation mediates behaviour in the labour market and how locational decisions affect both socially disadvantaged, highly segregated and marginal communities. Directive policy and additional equality practices in the labour market and creating more jobs in places of disadvantage have unquestionable merit. However, greater subtlety is required about how these ideas are worked out in practice, and specifically how land use planning can create arenas for sharing and mixing, especially in sectors of the growth economy.

MEASUREMENT PROBLEMS

The religious monitoring of workplaces has indicated that during the 1990s there was a more equitable share of employment by religion, which was accompanied by the creation of more companies that were less likely to be dominated by one particular ethno-sectarian group. Evidence presented on a yearly basis by the ECNI, which measures and monitors the religious profile of private companies and the public sector, has shown that since the mid-1990s there has been a decline in the overall level of religious segregation within the monitored labour market. This, it is argued, results from increased mixing, employment decline within highly segregated companies, equality legislation and a reduction in workplace choice being predicated upon fear, prejudice and ethno-sectarian competition (Osborne, 1996, 2003). As noted by Joan Harbison (2000), the former Chief Commissioner of the ECNI: 'The Annual Monitoring Reports also show that there has been a reduction in the degree of segregation in the workplace. Companies with an under-representation of either Catholics or Protestants have both shown improvement' (2000:1).

As shown in Table 6.1a, 64.9 per cent of all employees monitored in 2003 by the ECNI worked in companies within which at least 60 per cent of all employees were either Catholics or Protestants. Table 6.1b indicates that 31.3 per cent of all Catholic employees worked in places that were predominantly Protestant, compared to 7.6 per cent of all Protestants who worked in predominantly Catholic workplaces (Table 6.1c). Nearly two-thirds of all employees in companies with segregation levels over 80 per cent were Protestants compared to around 30 per cent who were Catholics, a situation that reflected the high level of Protestant employment in workplaces within the Belfast, South Antrim and North Down arenas.

The data contained in Tables 6.1a, 6.1b, and 6.1c reflect the impact of employment levels within regions of Northern Ireland that are dominated by one or other community. This is not problematic, in a general sense, as one would expect travel-to-work areas to conform to the religious profile of the communities within which sites of production are located. A problem arises in places, such as Belfast, where the travel-to-work area is relatively uniform with regard to

religion and distance to workplaces but the employment profile of
some companies is skewed towards one religious group.

Table 6.1a Percentage share by workplace segregation and religion in private sector
companies in Northern Ireland 2003

% share Catholic/Protestant	% Protestant share	% Catholic share	% share of all employment
90+	73.3	23.2	1.5
80–89.9	64.9	32.0	11.5
70–79.9	59.1	36.6	24.2
60–69.9	54.9	40.3	27.2
50–59.9	49.2	45.2	35.3

Source: Data supplied by ECNI

Table 6.2b Percentage share by workplace segregation and share of employment
within predominantly Protestant private sector companies in Northern Ireland 2003

Predominantly Protestant	% share of all Protestant employment	% share of all Catholic employment
90+	1.9	0.2
80–89.9%	12.7	3.1
70–79.9%	23.2	11.0
60–69.9	23.0	17.0
Total	60.8	31.3

Source: Data supplied by ECNI

Table 6.2c Percentage share by workplace segregation and share of employment
within predominantly Catholic private sector companies in Northern Ireland 2003

Predominantly Catholic	% share of all Protestant employment	% share of all Catholic employment
90+	0.04	0.7
80–89.9%	0.8	6.1
70–79.9%	2.7	11.2
60–69.9	4.0	10.4
Total	7.6	28.4

Source: Data supplied by ECNI

The central problem in terms of any interpretation of workplace
segregation is that it is based upon a monitoring process that has
constantly disguised the extent of religious segregation. The exact

nature of segregation within workplaces is obscured by the release of monitoring data, presented by the ECNI, which does not provide information on all workplaces. At present, information on each site operated by a multi-sited company is combined into a global figure for the parent company.

Many multi-sited companies operate workplaces that are either highly segregated or mixed, yet when data from each site are combined, for monitoring purposes, the company evaluated appears to be relatively mixed. The failure to measure and account for this internal segregation means that the presentation of monitoring data dilutes the extent of workplace segregation within Northern Ireland. As noted previously, the religious monitoring of workplaces has been interpreted as indicating that there are more mixed workplaces, even though Russell has argued that the religious monitoring of workplaces, undertaken by the ECNI, does not provide for a 'detailed examination of workplace segregation' and that the present measurement techniques are a 'rather crude proxy for segregation' (2004: 42). Despite such criticism, there has been no examination of how more effective means of analysis could be advanced in terms of measuring workplace-based segregation.

As a result of this lack of analytical endeavour, one of the most contentious issues within Northern Irish politics remains acknowledged but unaccounted for. This failure to provide a rigorous examination of workplace segregation permits the articulation of strong ethno-sectarian discourses on labour market segregation and discrimination to be advanced without any serious critique or challenge. Moreover, the presentation of data that suggests a shift towards a more shared labour market undermines the reality of continued division, chill factors and the impact of residential segregation upon labour market outcomes. There are several other key difficulties in terms of using the data supplied by the ECNI in terms of accurately gauging religious segregation and religious under-representation. These difficulties include:

- data collection on companies employing between 11 and 26 employees only beginning in 1992;
- the exclusion of data collected on companies that employ between 11 and 26 employees from the published list of monitored companies;
- the non-collection of data from companies that employ fewer than 11 employees;

- the non-submission of data on part-time employees until 2001;
- the non-listing of companies that employ fewer than ten Catholics or Protestants;
- the non-presentation of any information regarding the overall level of religious segregation;
- the exclusion of information on monitored and non-monitored companies that usually constitute a fifth of all private sector employment.

The ECNI argues that withholding data on companies that employ between 11 and 26 employees is based upon presenting an unbroken longitudinal analysis from the original monitored data set. This is peculiar given that the collection of data regarding part-time employees in 2001 was included within the monitoring statistics with no regard for the sanctity of the original data set. Furthermore, Osborne and Shuttleworth (2004) and Anderson and Shuttleworth (2002) have proposed that it is the small-scale employers, who do not fall within the published monitoring requirements, who are those most likely to be religiously segregated.

In relation to the data collected by the authors on 32,872 employees within private sector multi-sited companies in 2003, it was found that the gap between the official level of monitored segregation and the level of segregation across constituent sites was extensive. This sample included around one in eight of all monitored private sector jobs in Northern Ireland. Nearly two-thirds of all employment at site level had an over-representation of either group that was at least 12 per cent higher than the monitoring data for each company presented by the ECNI. Furthermore, 21.2 per cent of all employment at site level was highly segregated (over either 70 per cent Catholic or Protestant), and the under-representation of either community was at least 20 per cent beyond the religious composition of respective travel-to-work areas. In effect, the majority of sites analysed were highly segregated even though the monitoring data indicated that each of these companies were relatively mixed.

As indicated within this particular sample, the issue of internal segregation is exposed when employment at each site of employment is taken into consideration. There is no doubting, as has been argued by a series of commentators, that the measurement of segregation at site or workplace level by the ECNI would show that the private sector is more segregated than is presently reported. Moreover, given the limitation of the data provided by the ECNI, it is impossible to conclude whether sites of employment have become more or less segregated, even though it is fair to state that workplace mixing has increased. The fundamental problem is that ineffective measurement obscures the ability to study meaningful labour market shifts.

Official data sets are also limited in that they preclude information that permits analysis of important social and other place-specific factors. Data, for example, on the occupational classification of employees within each company are not available, and as such it is impossible to conclude if this is an important factor that influences workplace segregation. In wider political terms the employment, for example, of low-skilled Protestant employees from segregated Protestant communities in predominantly Catholic workplaces has wider political connotations than the employment of middle-income Protestants from less segregated places. Given that the majority of violence within Belfast was performed within the city's most segregated places and that people from such places are more likely to be victims of sectarian intimidation, it is important that the relationship between such places, standard occupational classification (SOC) and sites of employment is established with statistical rigour.

CORRECTLY MEASURING SEGREGATION AND MIXING

As indicated within the sample below of worksites in Belfast, the issue of workplace segregation is highly contingent upon the relationship between residential segregation, SOC and the religious composition of sites of employment. The composition of employment within the sites studied, which are highly segregated, is linked to wider processes of boundary maintenance between Catholics and Protestants from within highly segregated areas.

Data were obtained, in 2002, from over 40 private sector employers within Belfast. The overall sample was equivalent to 18 per cent of all employment within this sector in Belfast. There was a near-equal share of employment within sites that employed fewer than 100, between 101 and 250 and over 250 employees. Employers were asked to provide data on:

- the postcode address of employees;
- the employees' standard occupational classification.

Employers are unable to supply details on the religion of employees owing to issues of confidentiality. However, they did supply information on the religious composition of each site of employment. Employers also provided insights into issues relating to the religious composition of workplaces. The data provided made it possible to:

- determine the share of employees who came from predominantly Protestant or republican/nationalist areas. Highly segregated areas were denoted as those places within which over 90 per cent of the residents were either Catholics or Protestants;
- determine the share of employees from segregated areas by their SOC;
- analyse the share of employees who did not come from highly segregated areas and their respective SOC;
- measure the distance travelled to work by all employees.

The data presented are subdivided in terms of those companies that were predominantly Protestant and Catholic (over 70 per cent of all employees) and mixed workplaces within which the employment of one group was no higher than 55 per cent. There was a near-equal share of employment within these three sub-groups in relation to SOC. This provided the capacity to determine a strong relationship between segregation, job type and site location. In sum, 24.8 per cent of employment was based within the first three major SOC groups. The remaining 75.2 per cent of employment was located within SOC groups 4–9. A significant share of all employment (54.2 per cent) was located in major SOC groups 7 and 8.

Eighteen sites of employment employed a majority Protestant workforce. Within these companies, 72.1 per cent of all employees were Protestants compared with 21.8 per cent of all employment held by Catholics. An additional 6.1 per cent of all employees at these sites were cited as non-determined by religion. Within these sites 46.8 per cent of all employees came from highly segregated Protestant communities compared with 7.5 per cent of employees who came from highly segregated Catholic communities. The share of all employment among those from highly segregated Protestant communities was equivalent to 64.9 per cent of all Protestant employment. The share of employees from predominantly Catholic communities was equivalent to 34.4 per cent of all Catholic employment.

This disparity in the construction of employment in terms of the volume of segregation and the relationship to residential representation indicates that Catholics working within these sites were less likely than their Protestant counterparts to live within segregated communities. Additional evidence indicated that two-thirds of all Catholic employees at these sites were located in major SOC groups 1–3. This compared with a ratio of 1 in 4 among those determined as Protestants. Catholic employment within these sites tended toward a middle-income and professionally based employment structure. An explanation for this mismatch, in terms of workforce construction, was offered by a personnel manager:

It's true to say that most of our employees who are Catholics are in the offices and not on the shop floor. I suppose they feel safer in that type of environment. The usual thing about management and the higher-paid employees is that they are middle class. You know that the middle classes never mention religion. Well, never publicly anyway. I suppose on the shop floor that it's a lot different. More politicised and I suppose hard for us to monitor what's going on. Put it this way, the worst thing that could happen to you in the offices or R and D labs is that someone would throw a bagel at you. On the shop floor you could get a spanner in the teeth.

The relationship between segregation and employment was also related to distance travelled. Employees from predominantly republican/nationalist areas were more likely to travel longer distances to work than their counterparts from predominantly unionist/loyalist areas. The average distance travelled by those from predominantly unionist/loyalist areas, who were located within SOC groups 4–9, was 6.7 km. Among those from predominantly republican/nationalist areas and who were located within SOC groups 4–9 the distance travelled was nearly three times greater than that of their Protestant counterparts. This latter group tended to be composed of those who came from segregated Catholic communities outside the Belfast Urban area, whereas the former group tended to come from more localised and highly segregated communities within Belfast. Evidently, there was an under-representation of low SOC group employees from highly segregated nationalist/republican communities within Belfast and an over-representation of such employees from predominantly Protestant communities within the city. The drawing-in of employees from predominantly republican/nationalist areas outside of Belfast could be based upon a less conscious knowledge of Belfast's sectarian labour market or an ability to disguise ethno-sectarian connections. As noted by a personnel manager:

I suppose it's easier if you a Catholic working here if you are from outside Belfast. I suppose if you say you are from somewhere out in the countryside most of the more aggressive ones in here, who could do you harm, won't have a clue where your actually from – whether it's Catholic or Protestant. But if you were to say, 'Hey lads, I'm from Short Strand' you might be in trouble as they will know where you are from and what the politics are of where you are from.

Among those companies located within predominantly republican/ nationalist areas the share of all employment held by Catholics was 71.7 per cent. The share of employment held by Protestants and those registered as non-determined was 22.6 and 5.7 per cent respectively. Nearly 50 per cent of all employees came from predominantly nationalist/republican areas. This equated to 69.4 per cent of all Catholic employment. Around 90 per cent of all employees from segregated areas who were located in SOC groups 4–9 came from predominantly nationalist/republican areas. According to personnel managers, it was estimated that nearly two-thirds of all Protestant employees were employed in SOC groups 1–3. It was also observed that for every six employees from a predominantly republican/ nationalist area there was around one worker from a predominantly unionist/loyalist place.

The distance travelled by employees from segregated areas was also heavily influenced by community representation. The majority of employees from predominantly republican/nationalist areas travelled an average 4.1 km and 3.2 km respectively among those from SOC groups 1–3 and SOC groups 4–9. This indicated that the employees from predominantly republican/nationalist areas came from within a very tight labour market arena. Employees from predominantly unionist/loyalist areas came from a more enlarged labour market arena, and in turn most were drawn from beyond the Belfast area. As with the predominantly Protestant sample, the minority workforce, in this case Protestants, were drawn from a more extensive geographical area.

In terms of the predominantly Catholic and Protestant workplaces it is obvious that high levels of segregation produced similar employment profiles with regard to the types of areas from within which employees were drawn. Most employees from the minority group, in both segregated sub-groups, were drawn from the professional SOC groups and travelled longer distances to work than those from the majority ethno-sectarian group. The higher the level of segregation the fewer the share of employees, within the minority group, from highly segregated places. This latter finding means that the construction of employment within highly segregated sites reflects wider residential patterns and that skills and qualifications impact upon labour market choices. Evidently, the construction of segregated worksites mimics the nature and meaning of residential segregation. Such evidence furthers the argument that segregated habituation distorts consumptive and productive relationships. It also manifests a further form of segregation upon spatial and material outcomes and practices.

One of the companies within the sample, which was predominantly Protestant, had previously employed a mixed workforce. The closure of a road leading from predominantly republican/nationalist West Belfast led to a near 20 per cent decline in Catholic employment. In this instance, the road closure meant that some Catholic employees refused to use the alternative entrance that passed through a loyalist estate and in turn either sought employment elsewhere or took up work-related benefits. This was evidence that, in certain instances, sectarianised perception is linked to evident notions of spatialised security and access.

The third sub-set was constituted by sites of employment that were either located within mixed areas, such as the city centre, or in places that can be accessed from both predominantly republican/nationalist or unionist/loyalist communities. The share of Catholic and Protestant employment at these sites was 48.7 and 47.3 per cent respectively. There was a near-equal share of employment in relation to SOC and religion.

Unlike the segregated sites, employees from SOC groups 4–9 were drawn, in terms of distance, from similar travel-to-work catchments. Furthermore, and in comparison to the segregated sites, the relationship between employment and the distance travelled to work was more uniform with regard to religion, segregation and SOC. This is important as it indicates that workplace sharing can be based upon high levels of employment throughout all of the SOC groups. More importantly, employees are also drawn from highly segregated communities in similar numbers.

What appears with mixed workplaces are virtually dissimilar relationships with regard to religion, distance, SOC and levels of residential segregation. Mixed workplaces drew 73.5 per cent of all employees from segregated places. This compares with 56.3 per cent for the predominantly Protestant workplaces and 56.7 per cent for those sites that were predominantly Catholic. Companies that had religiously mixed workplaces were twice as likely as their segregated counterparts to have employees from within those areas that are officially designated as being deprived. Mixed workplaces also produced a more equitable share of employment in terms of SOC, segregation and distance travelled. These workplaces clearly go some way toward deconstructing the authority and power of apparent ethno-sectarian custom. As stated by Green *et al.*:

Different factors of limited mobility, lack of confidence and religious factors intertwine, in complex ways, to limit perceived opportunities and to provide a *post hoc* rationalisation of behaviour. In this way, a job that is located in, or close to an 'unsafe' area is 'inaccessible' – whether or not it is possible to travel there'. (2005: 321)

The evidence provided also suggests that the majority of those who make up the marginal group, in terms of religion/political opinion, in highly segregated workplaces are generally drawn from areas that are not highly segregated. In workplaces that are more mixed the share of those from highly segregated Protestant and Catholic places is more equal. This is an important finding in that it suggests that the religious/political composition of workplaces can be understood in terms of residential segregation as well as merely religious denomination/political opinion.

The material provided from the sites studied indicates how territorial borders can fixate and regulate mobility of flows and the capacity to undermine the impact of wider segregation patterns. It is evident that segregated sites of employment create space as a medium as powerful as any other form of social formation, an example of how space is produced and how the sociology of space remains important and influential.

7
Ethnic Poker:
Policy and the Divided City

William Neill has set out the history of policy in response to Belfast's changing ethnic fortunes since the late 1950s by identifying the institutionalisation of ethno-sectarian problems and the incapacity of planning and policy-making to develop alternative identities. According to Neill:

The general context for ethnic conflict management in Belfast at the beginning of the 21st century could be called 'ethnic poker', with both sides unrealistically upping the ante against the other … This makes it difficult to construct and support a common civic identity, which in the circumstances, despite a raft of legal measures dealing with rights and equality issues, remains underdeveloped and brittle. (Neill, 2004: 205)

The modernising project of 1960s Unionism, which attempted to create the conditions for inward investment and economic growth in the interest of a new mercantile class, unravelled in the face of economic recession and the outbreak of political violence at the end of that decade. Security imperatives took over and planning was preoccupied by managing conflict and protecting commercial centres from paramilitary attack. Place imagery and marketing Belfast as a modern, neutral and progressive city followed in land use planning, urban regeneration and social housing policy. Despite this shift Neill argued that planners are disconnected from the realities of 'ethnic poker' and either ignore or trivialise its consequences for urban management: 'The strategic urban planning response to the dual society that is Belfast in the year 2002 is a retreat to a position where the elephant is in the living room virtually ignored' (Neill, 2004: 214).

The state, politics and policy have become more conscious about segregation but its effects are read and interpreted selectively. While policy-makers are not in full retreat, their capacity to mount a coherent and effective response to the legacy of conflict has been muted. The policy landscape is now characterised by a scatter of unconnected initiatives, small-scale projects and funding streams aimed at a range of diverse problems. Each new urban crisis invites a new programme, often accompanied with a new layer of governance and a simply impenetrable maze of people, procedures and rules to understand and engage with. This chapter plots micro-programmes dealing with the specific interfaces, weak community capacity and environmental management, each with valuable effects but whose overall contribution is blighted by the lack of real policy imagination or vision. This is set against the legislative, institutional and resource investment in property capital and attempts to 'turn' Belfast's spatial economy and new consumerist landscape in such a way that it avoids the apparent realities of poverty and sectarianism. The problems of segregation are mediated via the prosaic administrative world of regional and local government, which in turn engenders a confused response ranging from denial and avoidance to the shifting of contentious issues to the next most 'convenient' body. There are, as shown here, some innovative projects that are linked to local consciousness-raising agendas, but these are rare and remain formative.

Pestieau and Wallace have made the point that 'ethno-cultural diversity is intersecting with city planning internationally over the same issues – community services and facilities, land use and zoning, economic development, architecture and urban design' (Pestieau and Wallace, 2003: 253). The restructuring of space in late capitalist cities, the withdrawal of the state from welfare provision and the need to construct new regimes to deliver planning outcomes have focused attention on urban governance, socio-economic and environmental problems and a concern for community interests (Alexander, 2002). As Belfast effectively lost its primary economy and had to discover new uses for old spaces, it too became more outward-looking to reposition itself in the global economy. The post-industrial, post-conflictual and post-bureaucratic transition of the region did refocus attention on exclusion, citizenship and, increasingly, the spatial legacy of communities left out in this progressive societal and economic shift. The preoccupation with reinventing Belfast as a modern, energetic European capital was witnessed in the creation of new consumption arenas in entertainment and leisure and increasingly in the residential landscape, which displays its own distinctive forms of social segregation. We saw in Chapter 4 that multiple segregations are affecting life chances and experiences and that social and ethnic space weave together in more complex ways to create a new geography of opportunity and denial, especially in places such as Belfast. As argued by Atkinson and Flint: 'While various "defended" territories exist in cities such as gang "turf", ethnic enclaves, gentrified neighbourhoods and areas of religious significance, gated communities provide a force for exclusion in new and different ways to earlier forms of residential patterning' (2004: 876).

The complex set of spatial relations has been empirically described in this book, and this chapter is concerned with the way that they are understood by policy-makers, how official discourses are constructed and operationalised and how contradictions emerge in the desire to maintain the modernising city project at the same time as managing violated and violent areas. It begins by explaining the treatment of division and violence within the city in 'territorial' policy arenas and draws attention to the interplay between urban crises, planning and professional values in determining policy outcomes.

This chapter uses specific examples from planning and urban regeneration to plot the struggle to adapt the policy agenda to these concerns, including the URBAN II Initiative and, more recently, the Development Corporation for the Laganside area, which has attempted to bring relevance to emerging policy priorities. This shift is an uncertain one and important gaps have emerged, especially in the delivery of resources and their impact on the socially excluded and segregated places. Policy has its own limitations to deliver change, and there is a danger in reducing the problems of space to the programme level and the implication that complex issues can be solved by careful or responsive planning. But, clearly it has a role to play in shaping spaces of opportunity, and this is where planning, as an instrument of urban management, needs to be more conscious of its distributive effects and reformist potential.

CRISES, BUREAUCRACY AND THE COLOUR-BLIND STATE

By the end of the 1950s it became clear that the structures and functions of the Stormont administration and local government were simply incapable of setting the region on a long-term development trajectory. A new business class had emerged within the old Unionist Party, and it valued economic and physical planning, especially in taking advantage of the increasing flow of global investment and 'branch-plant' economies. New economic plans and land use development strategies emerged in the early 1960s, as did a New Town Development Commission to organise the spatial economy, based primarily on international capital and manufacturing (Murtagh, 2002). Events overtook the tentative entry into interventionist government and accumulated into economic, social and, crucially, political crises, which in turn questioned the very existence of the Northern Ireland state. Crises were intimately connected to the production and use of land and the way in which resource allocations in housing, infrastructure and investment differentially affected the two communities. Modernisers within Unionism, seeking to secure primarily capital interests, were checked by reactionary instincts within the party and then steamrolled by the collapse of the 1960s civil rights campaign into intercommunal and ultimately paramilitary violence.

At the heart of these crises were economic stagnation, concentrated social deprivation and poverty, an inability to financially support major road infrastructure and inadequate urban redevelopment programmes. Other crucial factors included the poor condition and lack of supply of housing, the large-scale and rapid ethno-sectarian realignment of urban areas, civil disorder on a massive scale and the destruction of commercial premises as well as accusations that the military had interfered with planning and housing decisions to achieve security objectives (Murtagh, 2004). The response was not unlike that elsewhere in advanced capitalist states where organisational reactions to urban fiscal crises in the late 1970s and early 1980s had been to centralise expensive or contentious functions of government (Saunders, 1986). In Northern Ireland, planning, urban policy and housing were removed from local authority power and centralised within a new Department of the Environment Northern Ireland and in quango agencies, principally the Northern Ireland Housing Executive (NIHE).

Public participation in decision-making was restricted and highly corporatist forms of interest mediation were adopted to achieve planning outcomes (Neill, 1999). The legacy was a planning system where popular interests had little or no control and gained only selective access to the policy process and decision-takers. The Select Committee on Northern Ireland Affairs reviewed the planning system and concluded that 'running through the evidence was a feeling that in Northern Ireland people have little say about planning decisions' (House of Commons, 1996: 37). The use of highly regulated fora to debate and decide major planning outcomes has been criticised by Milton (1993) in her review of the public inquiry into the Belfast Urban Area Plan 1986–2001. She based her consideration of the transportation element of the plan on discourse analysis, which focused attention on the nature of communication and the language used in the inquiry. Milton found that the 'plan' was not the result of inclusive dialogue between different interests, mainly because the inquiry itself limited the scope and content of analysis.

In the field of housing, the Policy Studies Institute (PSI) report for the SACHR examined equality and access issues and highlighted the high rates of segregation in the public sector compared with other tenures (Smith and Chambers, 1989). Their study concluded that there was no evidence of direct or indirect discrimination by the NIHE, but Melaugh (1994) argued that the lack of policy and benign acceptance of segregation in tenant allocations and transfers encouraged this trend. The SACHR Report (1990) recognised the limitations on policy-makers to promote integrated housing and the pragmatic realism of that approach. Singleton (1986) also pointed out that while the NIHE did not have an explicit policy response to segregation, ethno-sectarian considerations were factored into routine planning and housing management decisions, especially at a local level.

The most comprehensive statement on the NIHE attitude to segregation, at that time, came in a paper prepared as part of a wider review of housing policy in Northern Ireland (NIHE, 1994). That analysis rehearsed the difficulties of promoting integrated housing and the need to protect freedom of tenant choice. In a sense, the very 'wickedness' of the problem explained the lack of a normative stance, and the paper concluded by calling for greater clarity and the offer of new departure for policy. But that was only part of the explanation, and for a fuller understanding of the policy deficit it is important to understand the 'restless search' for state legitimacy in a period of uncertainty in public policy in general and housing in particular.

The outbreak of civil disorder in Northern Ireland in 1969 followed a period of civil rights activity in which discrimination against Catholics in the allocation and supply of housing was identified as a major grievance (Birrell and Murie, 1986). However, Birrell and Murie also pointed out that lack of professional planning expertise, the small size of local authorities and the lack of finance were all part of the crises in housing at that time. In response, the NIHE was established as a comprehensive authority with ambitious development and renewal objectives. Integral to the success of this coping strategy was the de-politicisation of policy by removing any trace of sectarian inference from practices and procedures. Murtagh (2002) noted that planners and housing managers have adopted largely technocratic value stances and bureaucratic systems in response to violence and division and this has helped protect professionals and the state from criticisms of bias.

Any suggestion of a more explicit confrontation of issues of segregation and conflict raised concerns that this objectivity might be diluted or, worse, drag the planning and housing system into pre-1969 nadir of discrimination and abuse. Scott Bollens described this as greyness where colour matters and pointed out that there were significant costs especially given the widening of social and ethno-sectarian segregation and the avoidance of these factors as crucial planning issues (Bollens, 1999).

A POLICY TURN

Ultimately, the shift in general policy was created by both global and locally specific factors. It was pointed out earlier in the book that the failure of liberal economics and property market planning to trickle down to the most disadvantaged communities helped to refocus attention on communities, social objectives and new governance arrangements (Healey, 2002). In Britain, this was reflected in an explicit policy agenda on social exclusion and neighbourhood renewal. This also struck a cord with European concern in urban policy for longer-term programmes aimed at restructuring local economies and integrating marginal groups into the labour market (Murtagh and McKay, 2003). Race, hyper-segregation and the political instability of ethnic ghettos became mainstream policy concerns across the advanced capitalist world (Freeman and Hamilton, 2004).

Fear, terrorism and the politicisation of immigration, especially with the re-emergence of new right politics in Europe, gave added impetus to the issue and the need for a state response that centred on places and their particular ethnic and social character (Davis, 2005). Squires and Kubrin (2005) mapped the costs of racial segregation in the United States and emerged with a similar account to that presented in Chapter 6. In particular, they found that black neighbourhoods had higher rates of crime, unemployment, ill-health and poor educational attainment than other racially defined areas. They concluded thus:

A depressing feature of these developments is that many of these differences reflect policy decisions which, if not designed expressly to create disparate outcomes, have contributed to them nevertheless. The upside is that, if policy contributed to these problems, it is likely that it can help to ameliorate them as well. (2005: 56)

The boundaries of these possibilities (at least in a Northern Ireland context) are not fully understood nor realised, but they are there. Table 7.1 shows that the Agreement, Joint Declaration and Programme for Government all defined housing, spatial communities and regeneration as instruments of reform in the creation and maintenance of a shared society. The most recent policy on community relations makes more explicit than ever the government's impatience with segregation and the need for more aggressive integrative policies across government departments, and not least in developing opportunities for mixed living:

A shared society, which is our goal, is at ease with wide individual diversity, from which dynamism and vitality stem. It is held together by a willingness to engage in dialogue, on a basis of equality, and by a commitment to the common good – by a culture of tolerance. (OFMDFM, 2005: 9)

In order to achieve this vision, government departments produce triennial Action Plans. Specifically, the Department for Social Development (DSD), through the NIHE, 'will ensure that residence in a particular area should be a matter of housing need or personal

Table 7.1   Policy, politics and land

An essential aspect of the reconciliation process is the promotion of a culture of tolerance at every level of society, including initiatives to facilitate and encourage integrated education and mixed housing.
The Belfast Agreement, 1998: 17
The two Governments recognise that Northern Ireland remains a deeply divided society, with ingrained patterns of division that carry substantial human and financial costs. They recognise the importance of building trust and improving community relations, tackling sectarianism and addressing segregation including initiatives to facilitate and encourage integrated education and mixed housing
Joint Declaration by the British and Irish Governments, April 2003: 8
Growing as a Community Sub-priority 2: We will improve community relations and tackle the divisions in our society:
During 2003, put in place a coordinated plan of short term actions across government and key agencies to help build trust and confidence within and between communities including: action to support the capacity of local communities to deal with matters of dispute or division, including the proliferation of sectarian graffiti, unauthorised flag flying, the erection of memorials that can lead to community tensions; improving communication at interfaces; underpinning community development; and developing intervention programmes for young people.
Programme for Government, 2003: 18–19

choice rather than an insistence that only "one sort" live on certain streets or districts' (OFMDFM, 2005: 30). In pursuit of this the NIHE is directed to develop a number of practical actions including to:

- bring forward as soon as practicable its proposed pilot schemes on integrated housing;
- ensure that applicants wishing to live in areas where people of all backgrounds are welcome should have a real choice, subject to availability;
- consider how best, in consultation with PSNI and others, to protect mixed housing areas;
- continue, through the new Community Cohesion Unit, to support relationship building at neighbourhood level.

Part of the explanation of this shift also rests with the Northern Ireland Act (1998), which provided the legal interpretation of the Agreement and set down formidable equality duties on all public bodies in the region. We saw earlier that the Act established a Northern Ireland Human Rights Commission and introduced a statutory duty on public bodies to have due regard to the need to promote equality of opportunity for all categorised groups:

- between persons of different religious belief, political opinion, racial group, age, marital status or sexual orientation;
- between men and women generally;
- between persons with a disability and persons without;
- between persons with dependants and persons without (Section 75(1)).

A new spatial map of deprivation, the Noble Index, helped to direct urban and rural development programmes and in particular, the DSD's *Neighbourhood Renewal Strategy*, and *People and Place* strategy (DSD, 2003a). *People and Place* identified segregation as a significant component in understanding the multi-layered nature of urban disadvantage and the need to tackle the blighting effects of interfaces in wider regeneration programmes. The NIHE established a new Community Cohesion Unit, and the main physical development plan for Northern Ireland, the Regional Development Strategy (RDS), committed itself to the development of a shared and pluralist society.

IMPACT ON POLICY CONTENT AND DELIVERY

A leading planning theorist, Patsy Healey, was particularly impressed by the investment made in the preparation of the RDS and its capacity to reflect the delicate political context within which it was produced:

It is the outcome of a long and quite broad process of articulation. Its institutional ambition was primarily generative, to mobilize new ways of thinking about territory, with which to focus and justify investment. It sought to stabilize and build more territorial coherence in a divided society, build democratic opportunity, capture investment opportunity and create place qualities for both citizens and investors. (Healey, 2004: 64)

Table 7.2 is sourced from the RDS, and demonstrates a commitment to understanding the spatial effects of segregation and the need to address interface issues specifically. It also relies on a dual approach, in that it at one and the same time respects segregation while promoting integration where possible.

However, not all the Development Plans produced since the publication of the RDS have addressed these substantive themes. The largest strategy, which is for the Belfast Metropolitan Area Plan (DOENI, 2005a), has made an attempt to comply with the spirit of the equality legislation. One particular issue here is the role of the Equality Impact Assessment (EQIA) in considering the impact of the strategy, not least in the context of the twin and potentially contradictory objectives of the RDS. EQIAs draw heavily on Environmental Appraisal methodology and are comprised of a number of interlocking stages:

1. Defining the aims of the policy;
2. Gathering the available research data on equality and the plan or policy in question;
3. Assessment of the impacts on the nine statutory equality groups.

Consideration of:

- measures that might mitigate any adverse impact; and
- alternative policies that might better achieve the promotion of equality of opportunity;
- consultation on the findings of the draft Equality Impact Assessment;

Table 7.2   RDS and community cohesion

Strategic Planning Guidance (3) objective
To foster development which contributes to community relations, recognises cultural diversity and reduces socio-economic differentials within Northern Ireland
SCR 3.1 Foster patterns of development supporting community cohesion
facilitate the development of integrated communities where people wish to live together and to promote respect, encouragement and celebration of different traditions; and promote respect, encouragement and celebration of different traditions, and encourage communication and social intercourse in areas where communities are living apart.
SRC 3.2 Underpin the dual approach by fostering community interaction which could also contribute, over time and on the basis of choice, to greater community integration
Develop partnerships between public, private, voluntary and community sectors to facilitate community cooperation and involvement in securing social, economic and environmental objectives:

Facilitate the removal of existing physical barriers between communities, subject to local community agreement;
Support the development of 'shared places' accessible to all members of the community;
Revitalise the role of town centres and other common locations well served by public transport as focal points for shopping, services, employment, cultural and leisure activities for the whole community and promote the development of major employment/enterprise areas in locations which are accessible to all sections of the community;
Improve and develop public transport to assist in providing safe and equitable access to services, facilities and employment opportunities essential to the vitality of local communities;
Strengthen the network of local museums and heritage centres and arts centres with a special focus on understanding cultural diversity; and promote cultural diversity through the creation of opportunities in the creative industries associated with the arts.

Source: Based on DRD, 2001: 34–5

- decision by the public authority and publication of report on the results of the EQIA;
- monitoring for adverse impact in the future and publication of the results of such monitoring.

The EQIA for the Belfast Metropolitan Area Plan (BMAP) followed this methodology and found that the strategy did not discriminate intentionally or unintentionally against any of the nine categories. Some of the main conclusions of the EQIA were that:

- Housing zonings were located in areas where both Catholics and Protestants could access them, and the assessment found that the size of the zonings (in hectares) reflected the scale of population and housing demand in the future.

- In employment, there were even numbers of zonings in Protestant and Catholic areas but there was more hectares zoned in Protestant areas. This was especially the case for the Major Employment Locations proposed for the city. The dilemma for BMAP planners is what to do about this distribution especially in terms of the limited number of industrial sites, the emphasis on brownfield development and the commitment to locating new jobs in a smaller number of larger development sites.

Here, mitigating measures were important, in particular the extent to which public transport can connect segregated communities to growth opportunities in the urban economy. The EQIA concluded that the investment in rapid transit, new strategic transport corridors and the location of the 'Major Employment Locations' on strategic transport routes meant that they could be accessible to both populations without compromising their safety (DOENI, 2005b). Some initiatives have burned brightly, excited locals and then drifted or inexplicably expired. Special Task Forces were established in 2001 to report on the development of West Belfast and the Greater Shankill, and they set out 150 recommendations, linked mainly to job creation, training and inward investment. Some projects were supported by government, but not in any planned or integrated way, and the overall initiative was characterised by the same crisis 'feel' of other partially motivated interventions in area-based development of the city.

URBAN POLICY

Urban policy also made tentative steps to engage the effects of segregation and poverty, especially with the shift away from large-scale property capital projects and towards more community-led, neighbourhood-centred initiatives. The Department for Social Development (DSD) closely aligned its policies to Targeting Social Need (TSN) and its *Neighbourhood Renewal Strategy* and *People and Place* strategy highlight the connection between territory and urban development. For instance, it prioritises actions that will: 'Deal with the impact of the interfaces and peacelines on communities, including working with local people to explore how contested places might be better used' (DSD, 2003a: 25).

Three-year rolling strategies are designed to integrate four strands under the headings of physical, social, economic and community renewal, with connecting actions identified for a range of departments and agencies across government. The strategy built upon the experiences of EU-funded programmes including PEACE II and URBAN II, which piloted area-based interventions in highly contested places. With PEACE II, for example, the department is targeting its Measure 2.11 on interface areas and is working through local partnership structures to address the multiple needs of communities living on either side of interfaces (as well as some single-identity communities). Table 7.3 identifies the neighbourhoods targeted in the measure.

Table 7.3 PEACE II Neighbourhoods under Measure 2.11

Area	Neighbourhood
Belfast	Short Strand – Newtownards Road, Albertbridge Road, The Mount
	Lower Springfield, Cavendish and Beechmount – Ainsworth
	Donegall Road – Roden Street and Rockmount
	The Markets – Lower Ormeau and Donegall Pass
Londonderry	Brandywell – The Fountain
	Gobnascale – Irish Street
	Tullyalley – Curryneirin
Strabane	Fountain, Springhill Park
Downpatrick	Model Farm, New Model Farm and Flying Horse
Coleraine	Ballysally
Craigavon	Ardowen, Legahory, Burnside, Parkmore, Drumellen
Ballymena	Ballykeel

The EU URBAN II Programme reflected the broader European concern for the future of the city region and drew on pilot projects, the experience of Integrated Development Operations and the impact of URBAN I (Chorianopoulos, 2002). The programme has two objectives:

- to promote the formulation and implementation of particularly innovative strategies for sustainable economic and social regeneration of small and medium sized towns and cities or of distressed urban neighbourhoods in larger cities;
- to enhance and exchange knowledge and experience in relation to sustainable urban regeneration and development in the Community.

In Belfast, the Programme was targeted on inner North Belfast and is designed to invest €17.10m between 2000 and 2006 in a community of around 31,000 people. The selection of the area followed a period of extensive consultation, which highlighted the particularly intense nature of local problems. First, the area is stratified by six interfaces and by a long history of fear, sectarian deaths and internecine conflict. Second, North Belfast has suffered disproportionately from the effects of de-industrialisation, redevelopment and depopulation. The lost jobs in primary industries were not replaced by new employment in growth sectors of the economy. Low skills, low educational attainment and even basic literacy and numeracy skills reflected the area's wider decline, and the Noble Index showed that the most disadvantaged cluster of wards in Northern Ireland was concentrated in North Belfast (DSD, 2001). A significant component of the area identification process was that it was supported by the cross-party Social Development Committee and by the Democratic Unionist Party (DUP) Minister for Social Development, Nigel Dodds, before the subsequent suspension of Assembly business.

An area-based profile analysis set the context for the URBAN II strategy and highlighted the deeply structural nature of the area's economic demise. In particular, it identified the need to restructure redundant and blighted land and buildings and regenerate the property market especially in a way that met community needs and priorities. Key development opportunities were defined, not least because of the sheer quantity of derelict sites in the designated area. The analysis showed a need to restructure the labour market and promote realistic job chances for local people. This was especially important in any attempt to integrate young people and the long-term unemployed into the labour market. There was also the need to build the community infrastructure of the area, reduce tension and fear and facilitate a process of bottom-up renewal in North Belfast. Finally, the analysis highlighted the need for integrated planning in the area to maximise resources as well as specific technical skills to allow the local community and potential beneficiaries to engage with the programme effectively.

Following this analysis the strategy attempted to develop an integrated approach captured by the overall mission statement:

- to regenerate inner North Belfast into a vibrant, safe and viable urban community for its people, its environment and its economy.

Specifically, it developed operational priorities, which reflected the economic, environmental and social problems facing the North Belfast area:

- to develop the physical and social resources of the area;
- to develop the resources of local people especially to gain access to lasting employment opportunities;
- to develop and apply the technical competencies required ensuring that the Programme is implemented effectively and efficiently. (DSD, 2001: 40)

The Programme is being delivered by the North Belfast Partnership, which has drawn together representatives from the statutory, private and community sectors to manage the wider regeneration of this part of the city. The Partnership had struggled to hold together unionist/loyalist and republican/nationalist interests on one Board, especially given the violence around the Holy Cross dispute and the effects of the traditional marching season. But, intracommunity as well as intercommunity rivalries, including paramilitaries and organised crime, cut further into the contested nature of North Belfast. Allocating scarce financial resources in this environment is a delicate 'no-win' task, and the very fact that a Programme Management Executive has been established to deliver the measures and meet 'European Union Decommitment' targets on expenditure is an achievement in itself. Progress has been slow in dealing with interfaces and contested sites, but there seems to be little alternative to the slow iterative process of building confidence, some mutual trust and ultimately viable project ideas. As Amin noted: 'there is no formula here other than the engineering of endless tasks of interaction between adversaries or provision for individuals to broaden their horizons, because any intervention needs to work through, and is only meaningful in a situated dynamic' (2002: 969).

A number of flagship projects are being brought forward for implementation on or near the interfaces underpinned by painstaking work with communities on both sides. An obstacle to the more rapid delivery of the initiative has been the differential capacities of the unionist/loyalist and republican/nationalist communities in the area. Unionist/loyalist communities tend to have less experience, people and skills in local development while republican/nationalist communities have a stronger track record in negotiating their way through the grant regimes of Europe as well as mainstream regeneration programmes (Shirlow and Murtagh, 2004). Building 'binding' and 'bridging' social capital was established as a priority for the URBAN II, but what is less clear is what skills, at what level and in what sectors the investment should be made. The need for structured support in training, education and professional development in issues such as the social economy, understanding financial systems and project management is pressing if the initiative is not to pander to sectarian interests and resource competition.

Not surprisingly, much of the investment in area-based development has tended to concentrate on building the capacities of both, but especially the unionist/loyalist community, in order to engage resource-based programmes more effectively. The following section locates these efforts in policy thinking about the value of these capacities to policy delivery but also signals a potential downside in the specific context of Northern Ireland.

THE STATE AND THE CIVIC

Clearly, there has been progress in the attitude and response of the state to territoriality and the legacy of conflict. However, viewed in the context of the wider strategic development of Belfast in the last thirty years, a number of contradictions and inconsistencies emerge in the way in which problems are conceptualised, managed and resourced. For instance, while there has been a response within a number of sectoral areas concerned with the use and development of land, it is not clear who leads, where the strategic oversight rests and how different departments and agencies work together on very similar problems and opportunities.

This is in sharp contrast to the development of the Laganside area since the late 1980s. Here, a defined territorial problem was clearly mapped, analysed and planned for in the redundant docks and river front. A new body with strong legislative, financial and executive powers sought to regenerate the area with a long-term vision aimed at large scale investment, environmental improvement and job creation. We show later that it was successful in achieving these objectives, yet it begs the question why no equivalent response was created for interfaces across the city. The state's preoccupation with the use of capital to reimagine Belfast has produced a selective city capable of being used and enjoyed by those with resources, skills and knowledge. In turn, this has let the socially disadvantaged drift or more properly sink into an abandoned economic and ethno-sectarian landscape, especially in the inner city.

Moreover, a disconnection or implementation gap has emerged between high-level policies and delivery programmes across government. The commitment in the Regional Development Strategy to a pluralist environment has not found its way with any real vigour into local development plans. BMAP represented a significant opportunity to provide a creative vision for the city, move away from the regulatory land use approach and connect the agencies and resources vital to the effective management of ethno-sectarian space. Belfast City Council produced its own Master Plan (at the same time as BMAP), and this in turn set out an agenda for urban development but had no status outside the Council, attracted few resources and created confusion about policy ownership and city visioning. Strategy and governance competition characterises the void left by the lack of a clear approach to the comprehensive management and development of Belfast.

Unquestionably, part of the explanation for the gap between rhetoric and reality rests in the lack of skills and expertise both inside and outside government. There has not been a strong tradition of learning, reflexive practice and sharing ideas, and nor have the professional institutes in planning and housing clearly defined and supported the competencies needed for managing divided places. In the absence of a competent and creative steer within government, the struggle to manage crises at a local level has focused on civic society for answers. To some extent, this has tended both to isolate and internalise the problem, and while the learning is important, it is partial. The analysis and management of territorial problems require a broader context that understands the economic, social and environmental aspects of locality change and development. The investment made to develop the capacity of groups in order to turn their neighbourhood around, deliver an incredible pace of growth and heal divisions in some of the most polarised places in the city seems dangerously optimistic.

Policy-makers have invested in ideas about social capital and capacity building, which aim to activate localised social networks in order to mitigate the impact of social exclusion. However, after more than two decades of sustained funding in community infrastructure, especially through the Structural Funds, there are few empirical studies into its effects, not least in the delivery of area-based development programmes. Others doubt the reformist potential of social capital and generalist notions of capacity:

My first worry about 'capacity building', as about 'community development', is that it often seems to be a way of expecting groups of people who are poorly resourced to pull themselves up by their collective boot straps. So-called social capital is expected to take the place of economic capital. The imputed absence of social capital is potentially stigmatising and laid at the door of (mainly) poor people themselves. 'Capacity building' may be an alternative to economic regeneration. A large part of the effective resourcing therefore takes the form of unpaid work. (Levitas, 2000: 196)

Moreover, Ploger (2001: 236) argued that participatory practice is itself a form of corporatism, because participation is subordinated to formal structures as well as a presupposed political consensus. Raco (2000) pointed out that this is especially the case in urban regeneration programmes where some community interests have more to gain than others. The representative capacity, unitary conceptions of 'community' and the actual impact that community groups have remains under question especially with regard to the development of programmes that are linked to state legitimation processes (Shirlow and Murtagh, 2004).

Many of these dilemmas have been articulated in task force and policy reports, such as the 'Report of the Project Team' led by The Reverend Dunlop into violence in North Belfast set up in the aftermath of the Holy Cross school dispute. This recommended the following priorities for the area:

- develop an overall strategy for North Belfast;
- enable government to respond in a more joined-up way;
- boost community capacity;
- address interface issues;
- improve the economic, social and cultural life in North Belfast. (Project Team, 2002)

The North Belfast Community Action Project proposed a new capacity-building programme for the area involving a £3m five-year Community Empowerment Fund to develop weak community infrastructure. The Community Action Group (CAG) flowed from the North Belfast initiative and was set up to coordinate action across government and key agencies to help build trust and confidence within and between communities. The CAG had the following aims:

- improving communications at interfaces;
- supporting and assisting local communities in dealing with sectarian displays and emblems through local environmental improvements;
- underpinning community development, including streamlining arrangements for funding local community organisations;
- developing intervention programmes for young people; and
- generating debate and discussion on a new good community relations policy framework. (Project Team, 2002: point 155)

The CAG was also to act as a mechanism for partnership work involving government departments, public authorities and non-statutory organisations. It had a geographic focus on communities at interface areas with a high incidence or history of poor relations, those experiencing high levels of multiple deprivation and communities that had developed positive relations by supporting and encouraging good practice.

In real terms the whole initiative led to the establishment of a £3m Local Community Fund in February 2003 to be operated by the Department for Social Development. The aim of the fund was 'to empower people in disadvantaged communities, where community infrastructure is weak, to develop community capacity and leadership' (DSD, 2004: 1). Five objectives were specified for the fund:

- to develop community capacity and leadership;
- to promote partnership working, within and between communities;
- to help communities improve their local environments;
- to develop intervention programmes with young people; and
- to encourage more active participation by women in local community services. (DSD, 2003b: 1)

The first block of funding has now been allocated to local projects, with nearly one-third of the investment going to projects in Belfast. Launching the funding, the Minister, John Spellar, stated:

The Local Community Fund is all about giving local people the means to create a better future. It can allow them to take small but important actions such as improvements to the look of their street or estate or to create better local activities for their community both young and old. (DSD, 2003c)

What is interesting here is the way in which the problems of weak communities are understood and effectively internalised in terms of poor leadership and inadequate capacity. The assumption is that developing a culture of participatory practice draws the community away from paramilitary influence and sets it on a course of alternative development. Without being intimately connected to wider social and economic programmes, the growth economy and progressive political progress, it is difficult to see how this approach might move beyond its crisis feel. New partnerships and related participatory and coordinating structures lack the legislative weight, resources or skills to do the job of local development effectively. Fresh governance arrangements are grafted on to existing ones to create a bizarre and impenetrable delivery map that harms rather than facilitates the chances for integrated planning.

A particular form of community-led development identified by the Local Community Fund has focused on interventions in the environment, a common agreeable agenda around which both communities could negotiate with minimum risk (Neill, 2004). The 'Creating Common Ground Consortium' is a strategic partnership of seven statutory and voluntary agencies, plus two advisory members (Table 7.4). The consortium was appointed by the New Opportunities Fund in September 2000 to deliver the Green Spaces and Sustainable Communities programme up to 2006. The project concentrates on:

- 40 main housing estates in which multi-themed regeneration initiatives are led by the local community in partnership with technical support from the consortium;
- 10 estates with a specific focus on community safety;
- an Open Grant Programme whereby groups apply for funding for specific environmental projects.

The Housing Executive provides the secretariat for the programme, which had an initial grant of £5.28m, although the Housing Executive expects this to lever additional funding of between £15m and £20m. Estates have been selected on six criteria, which are: built environment; natural environment; community relations; community safety; community capacity; and social exclusion.

What is concerning is that, to date, there is little publicly produced evidence of the impact of these interventions beyond describing their organisation and activities. As highlighted earlier, there is a potential dark side of difference and negativity if this investment in community capacity does not consider the lasting effects on representational structures, the life chances of local people and the need for dialogue between investors and residents.

Table 7.4 Common Ground Consortium

Status	Organisation
Full member	Northern Ireland Housing Executive
	Community Relations Council
	Groundwork Northern Ireland
	Northern Ireland Office
	Community Foundation Northern Ireland
	Department for Social Development (Urban Regeneration)
	Department for Agriculture and Rural Development
Advisory members	Business in the Community
	Sports Council for Northern Ireland

There is also less clarity on what community capacity is for, what specific competencies are required and who will supply them. Ultimately, there is a danger that capacity building is limited to a number of procedural tasks that give the impression of serious devolution of control but which, in the end, add little to the lives of local people.

HOUSING AND THE ETHNIC MAP

One of the most interesting and significant policy shifts in the last ten years has been in the field of social housing policy. It was shown earlier that housing was one area where technical procedures and bureaucratic values had worked to completely renew the housing stock, meet massive accommodation need and ultimately remove a core grievance, especially among the Catholic community, about the discriminatory nature of the local state.

In 2005, the NIHE set up a Community Cohesion Unit and published an *Implementation Plan to Improve Community Relations Through Housing*. The aims of its good relations strategy commit the agency to:

- respond quickly and effectively to the needs of people in danger as a result of community conflict;
- work in partnership with others to address the complex housing needs of a divided society;

- respect the rights of people who choose to live in single-identity neighbourhoods; and
- facilitate and encourage mixed housing as far as is practicable, desirable and safe. (NIHE, 2005: 6)

There are interesting parallels here with the RDS, especially with regard to group rights and the link between identities and living patterns. The NIHE also aims to honour a rigorous commitment to maintain and promote its stock of mixed housing. Here, both ethnic and citizenship rights are recognised, although it is less clear what policy instruments will be used to achieve these disparate objectives, how they will be prioritised and how far the agency will expand the mixed housing debate into wider political debates.

At least within the field of housing there does seem to be some connection with programme delivery. Table 7.5 summarises relevant text from a range of area-based housing strategies. It is clear that there is analytical awareness of the implications of segregation, integration and territoriality for the delivery of housing plans in a wide range of contexts. The strategies specifically acknowledge the impact of interfaces, enclave communities, differential demographic characteristics on housing need and supply, blighted environments and stock conditions and the potential for integrated housing.

The North Belfast Strategy is especially interesting in that it explains the limits to housing management and construction policy and seeks to redefine the issue as a political and not exclusively a housing one. However, it also shows some organisational uncertainty about 'ownership' of the problem and whilst it implies a programme role for the agency, it asserts that leadership, regarding the wider effects of territoriality and segregation on the built environment, rests elsewhere.

TURNING LAGANSIDE

A number of organisations with land use responsibilities have recently confronted the problems of disadvantage and division. Yet it is clear that addressing the socio-spatial legacy of conflict will not rely only upon the internal machinations of divided places. The Report on 'Cities, Regions and Competitiveness' showed that the large English cities (apart from London) were at a distinct disadvantage against their Continental counterparts owing to underdeveloped knowledge-based economies, inadequate symbols of cosmopolitanism and culture,

Table 7.5 Spatial housing strategy and community divisions

Strategy	Quotes from the strategy documents
Shankill Sectoral Strategy	Greater Shankill contains four of Belfast's interfaces. These are at Crumlin Road, Springmartin Road, Cupar Way, and Lower Falls. The problems and issues surrounding the interface areas have been well documented. In housing terms conditions at the interfaces have required continued intervention. Vacant properties are common and turnover within social housing is high. Civil disturbance has forced many households to move from the areas. Dereliction is often a feature at interface areas and the poor environment has been improved in areas by landscaping and the encouragement of developments at Ainsworth.
West Belfast Sectoral Strategy	The Suffolk Estate is located on the eastern side of the Stewartstown Road around its junction with Black's Road. The estate was built in two phases during the 1950s and early 1960s. Problems affecting the estate developed as a direct consequence of the changing sectarian geography of West Belfast. Community polarisation transformed Suffolk into a lone Protestant area in West Belfast. The Board of the Housing Executive approved an Estate Strategy for the Suffolk Estate in 1986. By 1994, five phases of environmental and dwelling improvements were complete. However, voids continued to rise. The demolition of 94 dwellings was approved in 1994. A further review of the Estate in 1999 resulted in the demolition of a further 34 dwellings at Stewartstown Road in 2000 and their replacement with a Lidl supermarket in 2001. In addition a Suffolk/Lenadoon cross-community/commercial venture which received funding approval is currently being built on a neighbouring portion of cleared land.
North Belfast Housing Strategy	The housing problems in North Belfast are compounded by the deep-rooted sectarianism that remains a feature in some parts of the area and is graphically represented by the several 'peace lines'. There is no doubt that the segregated nature of North Belfast gets in the way of meeting housing need and prevents the best use being made of existing housing and land. While this cannot be right, in devising the Strategy we have had to recognise the practical limitations that segregation imposes. The Housing Executive cannot engineer territorial adjustments in North Belfast. In fact such a politically sensitive issue is beyond the ability of any agency to deliver.

social polarisation, environmental sustainability and fragmentation of governance and resource allocation (ODPM, 2003a). Sustainable cities need to be supported by broadly based development programmes that tackle both social exclusion and promote the growth economy.

Belfast's most successful attempt to spatially construct the growth economy was the Laganside Development Corporation. Its proponents saw it providing new opportunities for a city wrecked by violence and economic collapse (Smith and Alexander, 2001) but its critics criticised it as an exclusive Thatcherite property project that failed to connect with the city's disadvantaged and marginalised (Neill, 1995). Roger Tym and Partners (2003) conducted a meta-evaluation of English Development Corporations, which showed that they 'were successful at bringing about the physical regeneration of, and attracting investment to, areas suffering from market failure because of changing patterns of economic activity, degraded and polluted environments and fragmented land ownership' (Roger Tym and Partners, 2003: 8).

However, they were also critical of the limited budget for community development and showed that community support accounted for just 2 per cent of expenditure of all UDCs and no more than 5 per cent in any one organisation. Moreover, they found that pre-existing low-income households did not benefit from the work of the corporation; in short, the trickle-down effect did not happen. The government's latest response re-emphasised the functions of Urban Development Corporations around the 1980 Act and stated that their purpose would be to achieve regeneration through:

- bringing land and buildings into effective use;
- encouraging redevelopment of existing and new industry and commerce;
- creating an attractive environment; and
- ensuring that housing and social facilities are available to encourage people to live and work in the area. (ODPM, 2003b: 8)

However, in this version of the concept, the community would play a more central role in the design and delivery of regeneration programmes:

It is the Government's intention that whilst the UDCs' core role will be a catalyst for physical regeneration and development, the existing community will be at the heart of the UDCs' programmes. Working in partnership with existing agencies in social infrastructure and skills will be the key to the UDC delivering sustainable development. (ODPM, 2003b: 9)

Laganside Development Corporation (LDC) was established by statute in May 1989:

to secure the regeneration of 300 acres of land adjacent to the river Lagan by bringing land and buildings into effective use, encouraging public and private investment and the development of existing and new industry and commerce by creating an attractive environment and by ensuring that housing, social, recreational and cultural facilities are available to encourage people to live and work in the area. (DOENI, 1989: 20)

By 2000 the project had largely fulfilled its core aims in the development of the wider area, but criticisms of its failure to impact on the poorest people and areas remained (OECD, 2000). In particular, it was argued that the main outcomes related to the development of the property economy, the creation of highly skilled jobs and improvements in the public arena (Gaffikin and Morrissey, 2001). However, in line with other policy shifts, Laganside began to embrace the language of inclusivity and community in its more recent capital projects.

By 2004, a new Community Unit had been established, three community members appointed to the Laganside Board and an Equality Scheme prepared, which was supported by a range of Equality Impact Assessments on specific projects and programmes. Both TSN and Equality Plans were firmly located within a new Laganside Community Strategy, which aimed to integrate the Development Corporation's work with the adjacent 'Laganside local community' (LDC, 2003). A particular opportunity to translate these ideas into action came with the development of the Belfast Gasworks, which is an 11-hectare site released after the closure of the 150-year-old Town Gas industry in the early 1980s. The site was heavily contaminated and located in an inner-city area on the interface between the mainly Catholic Lower Ormeau and Markets areas and the mainly Protestant Donegall Pass communities. The site was additional to the original boundary of the LDC boundary, and because of Belfast City Council (BCC) ownership it has always had a high degree of independence from the main Laganside project in terms of land assembly, governance and strategy development. The Council spent £10.19m on additional land acquisition and development of the site and established the Gasworks Trust consisting of BCC, Laganside Corporation and local community groups to agree a development strategy for the wider area. It was agreed in the plan that the project should involve a mixed development with an emphasis placed on the creation of jobs suited to the particular labour market circumstances of the communities adjacent to the development site.

The first phase of the project included the reclamation of the heavily contaminated site, which involved progressive excavation of polluted soil. The second phase involved the creation of a quality environment through landscaping and the culverting of the Blackstaff River, which was a heavily polluted watercourse that traversed the site. New cycle and pedestrian paths were built, which connected the area to both the city centre and to an existing urban cycleway along the river. Development plots were released to the market on the basis of criteria that included the number and types of jobs being offered, design quality and the track record and financial security of the sponsoring developer. The City Council imposed an annual rack rent (of 10 per cent per annum on rent charged) on all developers, which has created a long-term income stream for reinvestment in the job aspects of the project.

In the initial phase this has allowed the Council to subsidise a local development agency to build 28 units for start-up businesses on the site. The Halifax Group built a major call centre, creating 1,000 jobs at a mix of skills levels and abilities; and the leisure group, Radisson, has recently completed a major hotel on the north side of the site. The most recent estimates suggest that when the scheme has been completed it will generate 657,065 sq. ft of commercial space at a cost of nearly £113m.

Table 7.6 shows that full occupancy is likely to generate a rental income of more than £6m, which, under the rack rent scheme will generate an annual development fund of £413,500 for reinvestment by the Council. The Council also suggested that the project has generated £2,444,000 from rates and that this figure is likely to rise to £2,918,000 on completion of the project. The jobs created in the development currently stand at 3,600, a figure that is likely to grow to 4,600 jobs on the completion of all projects.

Table 7.6   Completed developments and costs to 2005

Area sq. ft	Construction costs	Full occupancy rentals per annum	Indicator
256 000	£19 018 000	£1 341 700	1995[1]
343 200	£33 614 000	£3 216 000	1998[2]
540 065	£95 275 000	£6 094 500	Current achieved[3]
117 000	£17 750 000	£1 180 000	Pending development[3]
657 065	£113 025 000	£7 274 500	Final projected totals[3]

Sources: 1 KPMG study; 2 Colin Stutt Consulting; 3 Belfast City Council

The Gasworks Employment Matching Service (GEMS) was started by the Council in 1999 in response to concerns, raised by the Trust, that local people would not benefit from the development and the jobs it has created. The scheme operates as an intermediary labour market initiative, which links unemployed people with emerging job and training opportunities in the area. The initiative has the following aims:

- to provide a one-stop shop for employment support;
- to provide a co-ordinated and effective network of employment support services, which meet the needs of local unemployed people in order to return to employment;
- to provide specifically targeted information, guidance training and personal development support for people who are long-term unemployed;
- to be proactive in designing and managing outreach, training and personal development programmes that provide a bespoke response to the needs of the long-term unemployed people in the area, especially those identified as socially and economically excluded.

The service is particularly aimed at young people, the long-term unemployed, lone parents, people with disabilities, early school leavers and women returnees to the labour market. It is centred on six of the government's New Targeting Social Need wards, namely Blackstaff, Botanic, Shaftesbury, Ballymacarrett, The Mount and Woodstock. The service provides career guidance education, job training and practical help in applying for specific jobs via a job matching service. GEMS is delivered by the South Belfast Partnership but has a multi-agency Advisory Committee that includes representatives from statutory agencies, the community sector and business interests. The core outputs from the initiative are:

- a combined investment in the initiative of £774,000;
- 430 registered individual clients;
- 148 jobs achieved (7.5 per cent of all jobs created on the site currently);
- 208 completing training programmes;
- two jobs fairs, which attracted 2,500 people and 81 companies and had a total of 1,000 jobs on offer. (BCC, 2004)

The Gasworks case demonstrates the capacity of a large-scale capital project to confront issues of poverty and exclusion. Housing, planning and urban programmes are similarly turning their attention toward ethno-sectarian segregation and the impact that this has had on deepening the exclusion of some communities. What this might mean for programme delivery, the skills of professionals, the urban disadvantaged or Belfast's economic repositioning remains to be seen. In his analysis of regeneration in the city, Tyler argued that integrated action was required in three core areas, including strengthening the economic base, raising the competitiveness of the labour force and improving the public environment (Tyler, 2004). The spatial fracturing of the city along ethno-sectarian, social and economic lines severely limits the capacity to deliver any, never mind all, of these priorities in anything approaching an integrated manner.

Land use policy is in a period of transition. It is wrong to assume that planning, housing management and urban policy are blind to the realities of segregation and its effects on the city. Globally, late-capitalist states are struggling to understand and direct unpredictable urban change. New and more vicious forms of exclusion, which see poverty and race connect in more concentrated places, have challenged the scope of the policy agenda and the way in which programmes are delivered in a range of countries. In Northern Ireland, the peace process, equality and the potential to deal with the spatial effects of thirty years of violence and economic change have had specific effects. The Regional Development Strategy, BMAP, housing for community cohesion and Neighbourhood Renewal make rhetorical commitments to the promotion of spatial mixing and ameliorating the impact of segregation. This is a positive and encouraging departure from the proceduralist sterility of the previous thirty years, but it is less clear how policy will be put to work in specific situations.

Reconciling the spatialisation of the growth economy with the needs of marginal places will require strategic planning at the macro metropolitan scale. The evolution of Laganside and the Gasworks project has shown that connections can be made, but the narrow scope of regulatory planning, the compartmentalisation of government and weak local authority structures have blunted the potential of this type of integrated thinking. Michael Parkinson has commented: 'Governance and decision making is fractured and inefficient. There are a plethora of strategies for different parts of Belfast. But there is much less indication that these are capable of actually being delivered in a joined-up manner' (2004: 7).

The issue of segregation is falling between the cracks. There is a reluctance to lead, ownership is uncertain and is passed around risk-adverse agencies and professional disciplines. Local groups lack support and a coherent agenda to work to, there is insufficient investment in the skills of professionals to cope with this complexity, and knowledge and practice is poorly developed and shared. While demanding a more interventionist role from government, there is little evidence that after thirty years of support, community relations practitioners are offering policy-makers the support, advice or models that might realistically achieve it. Investment in community development, capacities and infrastructure is crucial, but it is not clear what the vision is, what it is for or even what specific competencies need to be acquired or deepened at local level.

One of the strengths of the urban debate in Britain, including the recent Task Force on Urban Renaissance, the Urban White Paper and the Policy Action Team reports on Neighbourhood Renewal, was the investment in understanding urban change, projective analysis and the emphasis on the potential of comprehensive planning (ODPM, 2003c). Integrated planning aimed at the creation of sustainable urban futures and linking policy to priorities of growth and social inclusion is crucial. Belfast's politics and trust are uncertain and volatile variables at the turn of the new millennium. Policy is not the answer to that volatility, but it can debate and help direct choices, it can shape opportunities and make clear the values that underpin progress to a genuinely shared society. The need to move beyond useful, but ultimately, rhetorical commitments to street-level reality is an urgent project both inside and outside government.

8

Conclusion

A central and increasingly important aspect in the analysis of place has been a comprehension of how identity is both represented and influenced by tiers of experience and the acceptance/rejection of pluralist discourses. Such analyses have examined how the perpetual fragmentation of place into multiple entities has come about owing to the representation and reproduction of multifarious dimensions of political and cultural resistance (Douglas and Shirlow, 1998; Gold and Reville, 2003). As argued within this book, the geography of segregation is associated with established as well as novel spatial politics and a desire, by some, to halt the dilution of place-centred and territorialised identities.

Ultimately, territorial conflict in Northern Ireland is played out via place-centred interpretations of struggle and is linked to discourses of torment, anguish, moral rectitude and the entitlement to challenge any 'terrorisation' and 'threats' that could occur. However, there is an evident gap, which has hopefully been challenged within this book, in terms of the volume of analysis committed to understanding the intracommunity and other philosophical differences that exist within highly politicised places. This lacuna in the research dedicated to place is peculiar, given that the ability to identify and ultimately mobilise those who are less influenced by notions of ethno-sectarian identity is crucial in terms of developing pluralist and less subjective categorisations of belonging.

Despite the observations concerning heterogeneity it has also been shown that the control of place is not merely a constant feature of intercommunity discord but also of intracommunity disunity. The politics of group identity are strong but not necessarily as pervasive as may be imagined. The strength of those political entrepreneurs who promote political identity around discourses of national identity and moral tales of suffering is based upon their capacity to encourage those whose lifestyle and identity is pluralist and secular to generally remain silent or to become apolitical.

The 'hardening up' of sections of the electorate has come about through the reality that traditional identities are threatened by forces beyond the control of localised leaderships. The knee-jerk reaction, which has plagued peace-building efforts, is a retort to change and a key instrument in the camouflaging of concessions. There are few signposts that point in an alternative direction – a common affliction within societies in which ethno-sectarian identification influences many at an early age – and this is reinforced by a series of institutional and community-centred instruments.

The types of lifestyles and identities located within segregated communities have generally been ignored within academic deliberation. Such groups are generally regarded as reactionary and overtly violent. Ultimately, there is a rejection of them and their ideology and representation. In a sense they do not display the 'rationality' that academics require to provide corroboration to them or to espouse their cause. Only groups with a 'progressive' perspective, such as those who promote bourgeois notions of community relations, have tended to be deemed worthy of examination.

There are very obvious reasons to study 'regressive' ethno-sectarian discourses and their impact upon society. First, meaningful differences exist within and between groups. Second, the differences between unionism/loyalism and republicanism/nationalism are perpetuated through the art and manipulation of territoriality. Third, the success of any group's 'leadership' is based upon the capacity to engineer the control or semi-permanent control of place as a platform from which to launch the commitment to universal ideas and beliefs. The marshalling of group identity within a political context is centred not upon the rendition of ideas but also through the practice of difference. Fourth, ethno-sectarianism, despite its obvious ideological content, is based not merely upon a struggle between peoples but also through a desire to shape and direct political and cultural transformation. The implication for the ideological production of territory in a place already saturated with various territorial divides is enormous. The promotion and reproduction of the segregated spaces, studied within this book, highlights how citizens tend towards the sealing of ideas via ideological enclosure. These closings are clearly dangerous impediments with regard to a fragile peace.

In recent times the choice of 'safe' groups to study has led to a failure to appreciate that territorial disagreements are accompanied by forms of spatial confinement, closure and violence. There is also a general failing within academic analysis with regard to misunderstanding the role and designation of peace-builders. It is usually assumed that the educated and 'rational' will play a significant role in conflict alteration. Academics, it is even assumed, will pour oil on troubled waters. It is hoped that human rights activists, womens' groups and community associations will develop the 'cosmopolitan' and 'sophisticated' politics needed for change. In deep and complex conflicts the capacity of paramilitary groups to assume forms of territorial control means that they are crucial to any process of conflict transformation. In terms of seeking progressive and/or pluralist voices, paramilitaries should not be immediately withdrawn from consideration. This is made all the more obvious given that potential civic leaders, from within the middle classes, walked away from the conflict in a desire to remain uncontaminated by it. They developed an art and fiction of blamelessness, which they hoped would absolve them from responsibility. The conflict that is Northern Ireland will only be resolved when the central protagonists shift in a forward and inclusive direction. At present they do so, but with caution and complex irony.

DIFFERENT, BUT NOT DISSIMILAR

The material on segregation and separation presented within this book is far from unique given that in many divided societies fear, prejudice, faltering peace-building strategies and ethnic chauvinism displace the construction and delivery of democracy. Ultimately, a fundamental and societal problem is the capacity or desire to deliver democracy, accountability and well-being concurrent with the idea of citizenship. Division in whatever form is exacerbated by the uneven distribution of power and opportunity, and a perpetual failure to recognise the spatial burdens placed upon those living in conflictual arenas.

The desire to locate differentiating marks between peoples is also problematic given that the reproduction of group loyalty is based upon the rejection of alternative political and cultural discourses. The desire of those who passionately wish to maintain group-based identities is to develop propaganda-centred perspectives that undermine the decency, moral worth and value of those to whom they are opposed. Introspection and self-centred acclaim is ultimately an undemocratic act. Those who vociferously promote group-based identities suffer from the fallacy that they are upholding meritorious and unchallengeable identities. In fact, such actions and deeds merely reflect an inept reading of diversity and a denial of cultural hybridity. The naivety of those who promote group-centred identities is that a great deal of what they believe in is generally constituted by the existence of something that they oppose. Unfortunately, in Northern Ireland the delineation of one community from the 'other' is an active and ongoing part of identity formation.

Essentially, segregation is reproduced not by the boundaries between places but through the deeds and actions that maintain the need and desire to remain separate. Somewhat depressingly, and despite the decline in violence, identity formation remains influenced by a real and imagined presence of an ontological 'other' that is 'threatening'. Such a depressing conclusion does not deny that sectarian consciousness is spread across a wide plain ranging from the mild and relatively benign acknowledgement of difference to the exclusivist and dramatic desire to remain uncontaminated by the presence or contact with the 'other'. The practical spatial relationship that is generally defined within this book is one based upon the desire to create 'good' relationships, between communities, by remaining apart. This benign 'solution' is at times rational given that threats abound, but such threats are not representative of a whole group, rather of sections within it. In effect, sectarianism is not merely a repressive relationship between communities but also within them, given that highly vocal sectarians undermine the capacity or desire to publicly articulate a shared intracommunity future. This problem is exacerbated when political leaders aim to define, to the group that they oppose, what their interpretation of democracy is. In so doing the appearance of intercommunal violence between segregated places has returned with gusto. This is not merely a matter of Irish and British identities requiring friction to survive but because there is 'no' alternative capable of challenging the sterile and repetitive politics of ethno-sectarian-based resource competition.

'Belfast is different' in that social, cultural and political divisions are so firmly attached to vociferous and rigorous ethno-sectarian discourses. However, the crucial point is that economic, social and cultural divisions that are located in other western societies are also (as is the case in Belfast) exacerbated by the actuality that democracy encompasses multiple and disagreed interpretations. It is not that Belfast is unique but that the passion and commitment attached to alternative perspectives and viewpoints is so endurable that it has the power to subvert the capacity to imagine non-sectarian solutions towards achieving conflict transformation. Sectarianism, for some, endures like a comfort blanket. An obsession with the 'other' community and the surveillance and ultimate counteraction of its deeds and beliefs remains as a central point for cultural and group cohesion.

Democracy is denied in other western cities, but the visibility of that abuse is generally less obvious than is the case in the concentrated and highly politicised environment that is Belfast. In Belfast, the performance of abuse, contestation and demand is undertaken publicly and with volume. The reason that such spectacle and violent abuse is apparent is due to the power of territorialised difference. Spatial separation provides some with a *raison d'être* to commit to action via highly politicised identities. Belfast is unique in that the disadvantaged do not alienate themselves by non-voting. That generally remains the preserve of a middle class who, bolstered by state expenditure, have removed themselves from civic responsibility and leadership. The motivation of class discontent is not framed by a Marxian notion of class alienation but by a desire to undermine the ethno-sectarian 'other' in the cause of 'protecting' the ethno-sectarian self.

It would appear that in the rest of the UK, as well as in the Republic of Ireland, the notion of national identity, once so crucially important, is declining as the idea of sacrosanct nationhood is diluted by mass cultures and the impact of other global forces. The middle classes, once the mainstay of Britishness, have chosen an alternative identity based upon second homes, home improvement and holidays in the sun. Of course, the rise of the British National Party and the UK Independence Party and the fears generated by Islamist militancy are a reminder that a classless and nationally emotionless UK is far from delivered. In the Republic of Ireland, the economic forces of consumer growth have aided the sweeping away of the praxis between church and state. But yet again the rise of racism and the conservatism directed towards the issue of abortion are reminders of a far from post-nationalist and utopian present.

Lifestyle may have changed but it is crucially important that critique is not mutated into self-referential discourse within which academics merely write about what they see, from the confines of the academy, as the world. The basics of poverty, racism, fear, prejudice and social fatalism abound, and there remains a requirement upon academics to shine a torch-light on those places in society where the monthly trip to IKEA, with a *latte* at lunch, remains an aspirant dream. Class, despite social mobility, remains a crucial category through which to explore life chances, motivation and belief. The appearance of rioting in the northern towns of England pinpoints the importance of fear, prejudice and socio-economic exclusion upon the maintenance of place. Such violence is far from Enoch Powell's 'rivers of blood', but important nonetheless. The use of place-centred imagery by Islamist extremists in the UK, in which Britain is envisaged as ungodly, racist and anti-Islamist, is centred upon the utilisation of depictions of 'self' and the 'other' in order to rationalise violent intent and in so doing forge a separation between the 'pure' and the 'uncontaminated'. It is a process that is akin to the most exclusive depictions of ethno-sectarianism in Northern Ireland.

Spatial inequality, in Belfast as elsewhere, is reproduced by the way in which segregated patterns are overlain on the geography of social need. It is simply not possible to treat these territories as unproblematic or reducible to narrow technical decisions. The construction of space and the way it shapes opportunity and denial and conceptions of 'otherness' is also relevant not only to Belfast but to racialised areas in urban Britain. The Cantle Report (2001) and Home Office response (2001) to the riots in northern British cities was critical of parallel lives, identities and schooling, and called for policy-makers to prioritise community cohesion in their treatment of segregated neighbourhoods. But the rhetoric of integration belies the power of segregation and the intricate geometry of fear, insecurity and place that explains the ontological value of clustering against the 'other'. At the same time, the processing of space restricts the 'opportunity set' available to the in-group. Space, as a variable for understanding injustice, has value beyond Belfast and the increasingly global flow of resources and knowledge that seems to reduce the significance of place as an empirical category.

Equality legislation and social need policy contain the potential to radicalise space, produce understandings of the way in which inequality and exclusion interlock to deepen poverty and create new possibilities for local mobilisation around material concerns, such as access to work, in both unionist/loyalist and republican/nationalist areas. But equality and social need have been located in attempts by the state to 'steer' Northern Ireland through crises of legitimacy, violence and international repute. This has shaped a positivist policy response where the reformist potential of both has been blunted as they are mediated through a set of bureaucratic practices and regulatory systems. No matter how well-intentioned policy-makers have grappled with the intricacies of strategic policy and law, they have generally failed to grasp the subjective meaning of social action and how this is worked through everyday decisions to create denial and exclusion. Policy-making dependent on a constructionist view would allow scope for an understanding of how fear, segregation and territoriality create and reproduce decisions about where to work and how to get there. The legislative force of equality provides ample scope to interpret its provisions in different ways, and there is an imperative to move beyond valuable but highly procedural systems concerned with sifting, proofing and editing policy production. Policy consumption also deserves attention. In particular, this book has offered methodological alternatives to understanding the connection between inequality, disadvantage and spatial relations. This in turn invites a different perspective on evaluating Equality and Targeting Social Need, and points to the normative possibilities of their coupling in the particular geography of disadvantage and sectarianism in Belfast.

Labour market inequalities and social differentials have been couched in terms of discriminatory practices, but this review shows that the variable of space and how it regulates behaviour and restricts choice is crucial. The location of places of production, facilities and community services clearly affects quality of life and life chances of the most marginal people in the city. Planning policy, for instance, needs to factor segregation into land use decisions, zoning policy and strategic transport provision. We need a profession sensitised to real space and the way in which ethno-sectarian territory shapes movement and interaction patterns. This invites a shift in Equality and Targeting Social Need into mainstream policy discourses from its current form as a separate and narrow technical set of regulatory procedures. Space, as a starting point to see how equality and disadvantage fuse to deepen exclusion and how it can offer radical alternatives to sectarianism and economic restructuring, has implications beyond the narrow streets and at times narrow minds of Belfast.

The general context of ethnic conflict management (as opposed to conflict transformation) remains given that, as shown, the power of 'ethnic poker' remains the instrument through which resource competition is played out. As argued within this book, the construction of the Belfast Agreement provided for and legitimised the capacity of each ethno-sectarian bloc to quixotically raise the demands they made of each other. Earnest efforts and endeavours were tempered by the need to represent the electorate in ways that suggested that their political representatives would have to ensure that they did not lose. The cost of this was that the common civic culture needed to advance alternative and civically designed policies was undermined. Moreover, the use of threat and physical violence against public sector staff also militates against the capacity to deliver fair and innovative services.

Investment in community development, capacities and infrastructure is crucial but it is not clear what the vision is, what is it for or even what specific competencies need to be acquired or deepened at local level. The fundamental problem remains in that each policy idea is deconstructed by community and political leaders in order to see how their community will benefit or, even more dismally, how it will lose out. In such an environment of ethno-sectarian cognition it is doubtful that policy will move beyond the rhetorical in the manner hoped for. This is not just a lack of ideas or enthusiasm but because virtually every instrument of governance and political representation is devoted to an ethno-sectarian 'self'.

THE SEARCH FOR THE COMMON GROUND

For Morrow (1996) the search for the 'common ground' as an explicit political objective took a number of forms. The first was through politics and the various attempts to establish consensual structures in the 1970s and 1980s. The power-sharing executive of 1973–4 involved unionists and nationalists sharing executive positions along with a 'Council of Ireland', which established formal institutional links with the Republic. It was the proposed council that was most objectionable to Unionists and which ultimately led to the collapse of the Executive. However, the Anglo-Irish Agreement (1985), the Downing Street Declaration (1993) and the Framework Documents (1995) all enshrined the principles of cross-border cooperation in the search for a political settlement. The Belfast Agreement (1998) formalised the strategy of internal power-sharing, institutionalised cooperation with the Republic through new cross-border bodies and a new Council of the Isles to strengthen east–west relations.

A second strand in the search for the common ground was the promotion of equity, particularly in the labour market. Fair employment legislation was strengthened throughout the 1970s and 1980s, culminating in the establishment of the Fair Employment Commission in 1989, with its strong legislative powers to enforce equity of opportunity and treatment in the workforce. Morrow pointed out the danger of restoring employment imbalance in an economy with low or static growth where new jobs in one community can only be achieved by decreases in the other. Other sectors, particularly education, integration and mutual understanding, were also promoted and the Standing Advisory Commission on Human Rights was established to examine issues of justice and human rights. The Central Community Relations Unit (CCRU) was set up within the Northern Ireland Office to advise on community relations issues across a range of policy areas. Supporting political and policy toward accommodation was a raft of initiatives aimed at encouraging community contact and respect for cultural pluralism (Knox and Hughes, 1995). In 1990, a new Community Relations Council was established to support community and voluntary sector groups working to promote better relations between Protestants and Catholics in Northern Ireland and understanding of the 'others' cultural traditions.

However, as indicated within this book, the political middle ground has floundered in the face of the growth of ethno-sectarian chauvinism and the failure of those who are pluralist and secular to combine their attitudes into political strategies. The British state upheld middle-class lifestyles, and it would appear that as a result of this community has slowly ebbed away from political responsibility. Their fellow anti-sectarians within the working classes tend to remain quiet and have opted to negotiate life beyond their segregated life world.

As shown within COSE, the Protestant and Catholic middle classes have negotiated their way to the shared space of leafy suburbia. The benefits of social and spatial mobility, the alliance across the ethno-sectarian boundaries and the material benefits of a consumptive class have been worked out in new territorial relationships. The potential of these alliances both to create connections and withstand sectarian shocks has not been entrusted to the deprived of Belfast's sink estates, even though the central political project remains to 'encourage change in identities, interests, incentives for compromise and perhaps even allegiances' (Ruane and Todd, 2003: 68). It is hard to see the street politics of this agenda rolling out without some challenge to ethno-sectarian competitiveness and some acceptance of the need to share resources, ideas and ultimately spaces. Cox has analysed the problems in delivering the Agreement in practice, arguing that the process of achieving a negotiated settlement has been hampered by the failure to create 'a culture of sharing' (Cox, 2002: 167). Ultimately, we have taken the soft option, asserting either the impossible, or at best the problematic nature of the reconciliation project. The main success thus far with regard to state planning of division has been to create an emergent Catholic middle class who feel ambivalent about their identity: a policy similar to saving all of the swimmers at the expense of the non-swimmers as the boat sank. As argued by Graham and Shirlow, the complexities of division are still deep and labels are superficial:

This is a society in which multiple constructed geographies of lifestyles, class and welfare and complex renditions of the relationship between place and identity are apparently subsumed within a political disputation informed solely by conflicting attitudes to the Union combined with essentially obsolete constructs of national sovereignty. (Graham and Shirlow, 1998: 245)

Victimhood and memory are active factors in the reproduction of the past within the present. The ethno-sectarian 'elephant' will not forget the transgressive acts of 'others'. It is evident that the issue of loss and harm will remain a tool of political manipulation as long as civil society refuses to make mature arguments concerning the past. A fundamental problem is that political protagonists have mobilised harm in such a manner that they have stripped away the humanity of loss to the bare essentials of political usury. In politicising victimhood political entrepreneurs have made the dead take sides.

Undoubtedly, social capital is a process that aids the understanding of, among other things, tolerance of diversity, an ability to cope with new people and cooperate with non-familiar people and ultimately a capacity to see common interests. This is especially relevant to Northern Ireland where ethno-sectarian groups have assembled an impressive array of social capital to reinforce exclusivity around contact, and ignore the unpalatable or merely the different. It is vital to measure the quality of the social capital and not merely frequency of usage as quantitative measures need to be supported by attitudinal and opinion data on social distance, integration and attitudes to others. As eloquently argued by Porter, we must remain aware of the following:

Strong reconciliation may indeed seem too demanding in a society that is accustomed to cultural and political attitudes being shaped by habits of self-interest, suspicion and contempt of the other … The drift in my emphasis on cultivating fair interactions, searching out common ground and seeking inclusive belonging, encourages citizens in the North to believe that instead of slotting into a friend/enemy distinction based on squabbles about real estate, inter-traditional cooperation, if not indeed fraternity, is a better option. (2003: 267–8)

In many instances the political instability that still exists reflects the limitations of the current 'peace process' and the ability of devolution to substantially alter the nature of conflict. The central goal of the Irish and British states is to be seen to promote 'parity of esteem' and 'mutual consent' via the promotion of political structures that underline pluralism. How these modes of pluralism will remove the realities of economic and cultural sectarianism remains unaccounted for. However, as evidenced by the information presented within this book, sectarian actions are still in certain arenas more voluminous than constitutional and pluralist words.

Despite the cessation of most paramilitary violence we are left with a situation within which the creation of territorial division and rigidified ethno-sectarian communities means that fear and mistrust are still framed by a desire to create communal separation. Residential segregation still regulates ethno-sectarian animosity through complex spatial devices. More importantly, the capacity to reconstruct identity and political meaning is obviated by political actors who mobilise fear in order to strengthen unidimensional classifications of political belonging. Community-based self-representation assumes the form of a mythic reiteration of purity and self-preservation. As such, the potential to create intercommunity understandings of fear is, in terms of politics, marginalised by wider ethno-sectarian readings. Despite significant and important changes, it is evident that wider senses of powerlessness are responsible for the failure of intercommunity politics to emerge.

Notes

1. The origins and history of the dispute over the Drumcree church parade were in 1995 when the annual Orange parade to commemorate the Battle of the Somme was blocked by republican/nationalist residents in Portadown. The issue of the right to march along the traditional route remains unresolved. The Drumcree conflict has become a symbol for issues that are at the heart of the sense of identity of both main parts of the community in Northern Ireland. It has therefore come to have a significance that transcends the confines of Portadown. For Orange Order supporters, the Drumcree church parade has become a touchstone for civil and religious liberty – their right to demonstrate their faith and their culture by maintaining an age-old tradition. The dispute has also acquired a strong political overtone because nationalists believe the opposition to the parade is manipulated by Sinn Fein and designed to inflict a 'defeat' upon them. For many republicans/nationalists living along the parade route, impeding the march is tied to the principles of equality and parity of esteem. The Drumcree dispute has led to high levels of interface violence across Northern Ireland including the death in 1998 of the three Quinn brothers in Ballymoney.

    The Holy Cross dispute occurred in 2001 and 2002 in the Ardoyne area of Belfast and involved residents of a unionist/loyalist area picketing children and their parents at the local Catholic primary school during the daily walk to and from school. Incidents of verbal abuse and violence occurred at the pickets and there was widespread associated disorder throughout North Belfast for the duration of the dispute.

1   EVEN IN DEATH DO US STAY APART

1. The terms republican and nationalist are preferred to Catholic. Similarly, unionist and loyalist are preferred to Protestant. In both cases such descriptions are more valid and their use undermines the misconception that the conflict is about religion. Furthermore, the dominant political discourse within each area is referred to first.

3   INTERFACING, VIOLENCE AND WICKED PROBLEMS

1. The Noble Index is a multivariate measurement of deprivation in Northern Ireland.
2. 11 July is the night during which those who support the Orange Order host fires to celebrate 12 July. 15 August is the Feast of the Assumption, celebrated by Catholics and the Ancient Order of Hibernians.

4   BETWEEN SEGREGATED COMMUNITIES

1. Taigs is a derogatory term for Catholics.
2. Huns is a derogatory name for Protestants.
3. The communities, listed by the republican/nationalist places first, were Ardoyne/Upper Ardoyne, New Lodge/Tiger's Bay, Manor Street/Oldpark, Lenadoon/Suffolk, Whitewell/White City and St James'/Village.

6   WORKSPACES, SEGREGATION AND MIXING

1. The MacBride Principles – consisting of nine fair employment principles – are a corporate code of conduct for US companies doing business in Northern Ireland and have become the Congressional standard for all US aid to, or economic dealings with, Northern Ireland. The Principles do not call for quotas, reverse discrimination, divestment (the withdrawal of US companies from Northern Ireland) or disinvestment (the withdrawal of funds now invested in firms with operations in Northern Ireland). The Irish National Caucus positively encourages non-discriminatory US investment in Northern Ireland.
2. The Industrial Development Board was replaced by Invest Northern Ireland in 1999.
3. Direct Rule over Northern Ireland was undertaken in 1972 during the breakdown in civil order. Until 1998 Northern Ireland was directly administered by the Northern Ireland Office. The formation of the Northern Ireland Assembly led to the restoration of most decisions to the Assembly and its ministries. Direct Rule has been reapplied during the suspension of the Assembly.

References

Ackleson, J. (2000) 'Discourses of Territoriality and Identity on the U.S. Mexico Border', in D. Newman (ed.), *Boundaries and Globalization in a Postmodern World*. London: Frank Cass.

Adair, A., Berry, J., McGreal, W., Murtagh, B. and Paris, C. (2000) 'The local housing system in Craigavon N. Ireland: ethno-religious residential segregation, socio-tenurial polarisation and sub-markets', *Urban Studies*, 37: 1079–1092.

Agnew, J. (1993) 'Representing Space: Space, Scale and Culture', in J. Duncan and D. Ley (eds), *Place/Culture/Representation/Knowledge*. London: Routledge.

Alexander, E. (2002) 'The public interest in planning: from legitimation to substantive evaluation', *Planning Theory*, 1 (3): 226–249.

Alliance Party (no date) *Communal Divisions* (www.allianceparty.org/home/www/political/htdocs/index.php?id=19).

Amin, A. (2002) 'Ethnicity and the multicultural city: living with diversity', *Environment and Planning A*, 34 (1): 959–980.

Anderson, J. and O'Dowd, L. (1999) 'Borders, border regions and territoriality: contradictory meanings, changing significance', *Regional Studies*, 33 (7): 593–604.

Anderson, J. and Shuttleworth, I. (2002) 'Does fear of violence influence where people are prepared to work in Belfast?', *Labour Market Bulletin*, 16: 147–154.

Archer, M. (1995) *Realist Social Theory: The Morphogenetic Approach*. Cambridge: Cambridge University Press.

Ashley, R. (1987) 'The geopolitics of geopolitical space: toward a critical social theory of international politics', *Alternatives*, 12: 403–434.

Atkinson, J. and Flint, J. (2004) 'Fortress UK? Gated communities, the spatial revolt of the elites and time–space trajectories of segregation', *Housing Studies*, 19 (6): 875–892.

Aughey, A. (1989) *Under Siege: Ulster Unionism and the Anglo Irish Agreement*. Belfast: Blackstaff Press.

Aughey, A. (2005) *The Politics of Northern Ireland: Beyond the Belfast Agreement*. London: Routledge.

Bairner, P. and Shirlow, P. (2003) 'When leisure turns to fear: understanding the reproduction of ethno-sectarianism in Belfast', *Leisure Studies*, 22 (3): 247–276.

Barnett, W., Baron, J. and Stuart, T. (2001) 'Avenues of attainment: occupational demography and organizational careers in the Californian civil service', *American Journal of Sociology*, 106 (1): 175–187.

Barth, F. (1969) *Ethnic Groups and Boundaries: The Social Organization of Cultural Difference*. Boston: Little Brown.

Bauman, Z. (2004) *Wasted Lives: Modernity and its Outcasts*. Malden: Blackwell.

BCC (Belfast City Council) (2004) 'Gasworks Business Park Statistical Profile'. Unpublished briefing paper.

Belfast Interface Project (1999) *Inner East, Outer West: Addressing Conflict in Two Interface Areas*. Belfast: BIP.

Belfast Interface Project (2005) *A Policy Agenda for the Interface*. Belfast: BIP.

Bew, P., Patterson, H. and Teague, P. (1997) *Northern Ireland: Between War and Peace. The Political Future of Northern Ireland*. London: Scarecrow.

Bew, P., Patterson, H. and Teague, P. (2000) *Northern Ireland: Between War and Peace. The Political Future of Northern Ireland*. London: Scarecrow.

Bhasker, R. (1998) *Critical Realism*. London: Routledge.

Birrell, D. and Murie, A. (1986) *Policy and Government in Northern Ireland: Lessons of Devolution*. Dublin: Gill & Macmillan.

Boal, F. (1969) 'Territoriality in the Shankill–Falls Divide in Belfast', *Irish Geography*, 6 (1): 30–50.

Boal, F. (1976) 'Ethnic residential segregation', in D. Herbert and R. Johnston (eds), *Social Areas in Cities, Vol. 1*. Chichester: John Wiley.

Boal, F. (2000) *Ethnicity and Housing: Accommodating Differences*. Aldershot: Ashgate.

Boal, F. and Murray, R. (1977) 'A city in conflict', *Geographical Magazine*, 44: 364–371.

Bollens, S. (1999) *Urban Peace-Building in Divided Societies: Belfast and Johannesburg*. Boulder: Westview Press.

Borja, J. and Castells, M. (1997) *Local and Global. The Management of Cities in the Information Age*. London: Earthscan.

Borooah, V. (1996) 'Overview and Conclusions', in E. McLaughlin and P. Quirk (eds), *Policy Aspects of Employment Equality in Northern Ireland*. Belfast: SACHR.

Bourdieu, P. and Wacquant, L. (1992) *Invitation to Reflexive Sociology*. Chicago: University of Chicago Press.

Boyle, K. and Hadden, T. (1994) *Northern Ireland: The Choice*. Harmondsworth: Penguin.

Bradley, H. (1996) *Fractured Identities: Changing Patterns of Inequality*. Cambridge: Polity.

Breen, R. and Divine, P. (1999) 'Segmentation and Social Structure', in P. Mitchell and R. Wilford (eds), *Politics in Northern Ireland*. Boulder: Westview Press.

Brenner, N. (1999) 'Beyond state centrism? Space, territoriality, and geographical scale in globalisation studies', *Theory and Society*, 28: 39–78.

Brewer, J. (1992) 'Sectarianism and racism, and their parallels and differences', *Ethnic and Racial Studies*, 15: 352–364.

Bullen, P. and Onyx, J. (1998) *Measuring Social Capital in Five Communities in New South Wales Communities*. Lindfield: Management Alternatives Ltd.

Burton, F. (1978) *The Politics of Legitimacy: Struggles in a Belfast Community*. London: Routledge & Kegan Paul.

Byrne, S. and Irvin, C. (2001) 'Economic aid and policy making: building the peace dividend in Northern Ireland', *Policy and Politics*, 29 (4): 413–429.

Campbell, G. (2002) 'Discrimination statistics' (http://www.news.bbc.co.uk/1/hi/northern_ireland/3515210.stm).

Campbell, R. (1989) *The Hypocrisy of Irish Conflict*. Bedford: Moon Books.

Cantle, T. (2001) *Community Cohesion Review Team Report*. London: HMSO.

Chorianopoulos, I. (2002) 'Urban restructuring and governance: north–south differences in Europe and the EU URBAN Initiative', *Urban Studies*, 39 (3): 705–726.

Cohen, A. (1985) *The Symbolic Construction of Community*. Chichester: Ellis Horwood.

Colin Stutt Consulting (1998) *Evaluation of the Gasworks Development for Belfast City Council*. Belfast: Belfast City Council.

Compton, P. (1995) *Demographic Trends in Northern Ireland*. Belfast: NIEC.

Connolly, M. (1994) 'Public Administration and Expenditure Developments in Northern Ireland', in P. Jackson and M. Lavender (eds), *The Public Services Yearbook 1994*. London: Chapman Hall.

Connolly, P. (2003) *Ethical Principles for Researching Vulnerable Groups*. Belfast, OFMDFM.

Connolly, P. and Healy, J. (2003) 'The development of children's attitudes toward "The Troubles" in Northern Ireland', in O. Hargie and D. Dickson, *Researching the Troubles*. Edinburgh: Mainstream Publishing.

Cormack, R. and Osborne, R. (1994) 'The Evolution of a Catholic Middle Class', in A. Guelke (ed.), *New Perspectives on the Northern Ireland Conflict*. Aldershot: Avebury, pp. 65–85.

Coulter, C. (1999) 'The absence of class politics in Northern Ireland', *Capital and Class*, 69: 77–100.

Cox, E. (1999) 'Can social capital make societies more civil?' *Australian Planner*, 36 (2): 75–78.

Cox, H. (2002) 'Keeping Going: Beyond Good Friday', in M. Elliott, *The Long Road to Peace in Northern Ireland*. Liverpool: Liverpool University Press.

Crowder, K. (2000) 'The racial context of white mobility: an individual-level assessment of the White Flight hypothesis', *Social Science Research*, 29 (2): 223–256.

Darby, J. (1976) *Conflict in Northern Ireland: The Development of a Polarised Community*. Dublin: Gill & Macmillan.

Darby, J. (1986) *Intimidation and the Control of Conflict in Northern Ireland*. Dublin: Gill & Macmillan.

Darby, J. and Morris, G. (1974) *Intimidation in Housing*. Belfast: Northern Ireland Community Relations Commission.

Davis, D. (2005) 'Cities in global context: a brief intellectual history', *International Journal of Urban and Regional Research*, 29 (1): 92–109.

Delaney, D. (2005) *Territory: A Short Introduction*. Malden: Blackwell.

Department for Enterprise, Trade and Investment (DETI) (2002a) *Draft Equality Scheme*. Belfast: DETI.

Department for Enterprise, Trade and Investment (DETI) (2002b) *New Targeting Social Need Action Plan: Progress Report for 2001–2002*. Belfast: DETI.

Department for Regional Development (DRD) (2000) *Shaping Our Future: Response by the Department of Regional Development to the Report of the Independent Panel following the Public Examination*, Belfast: DRD.

Department for Enterprise, Trade and Investment (DETI) (2003) *The Labour Force Survey: Religion Report*. Belfast: DETI.

Department for Enterprise, Trade and Investment (DETI) (2005) *Economic Performance Report from the Department of Trade and Investment*. Belfast: DETI.

Department for Regional Development (DRD) (2001) *Shaping Our Future: Regional Development Strategy for Northern Ireland*. Belfast: DRD.

Department for Regional Development (DRD) (2003) *Belfast Metropolitan Transport Plan: Working Conference Papers*. Belfast: DRD.

Department for Social Development (DSD) (2001) *Operational Programme for the URBAN II Community Initiative Programme 2000–2006*: Belfast, DSD.

Department for Social Development (DSD) (2003a) *People and Place, A Strategy for Neighbourhood Renewal*. Belfast: DSD.

Department for Social Development (DSD) (2003b) *The Local Community Fund*. Belfast: DSD.

Department for Social Development (DSD) (2003c) *Spellar Allocates £2.7m for Left Behind Communities*. Belfast: DSD.

Department for Social Development (DSD) (2004) *Local Community Fund*. Belfast: DSD.

Department of the Environment (NI) (DOENI) (1989) *Belfast Urban Area Plan 2001*. Belfast: HMSO.

Department of the Environment (NI) (DOENI) (2005a) *Draft Belfast Metropolitan Area Plan*. Belfast: Planning Service.

Department of the Environment (NI) (DOENI) (2005b) *Equality Impact Assessment, Belfast Metropolitan Area Plan*. Belfast: Planning Service.

Doherty, P. and Poole, M. (1996) *Ethnic Residential Segregation in Belfast*. Coleraine: Centre for the Study of Conflict.

Donnan, H. and Wilson, T. (1999) *Borders: Frontiers of Identity, Nation and State*. Oxford: Berg.

Douglas, N. (1997) 'Political Structures, Social Interaction and Identity Change in Northern Ireland', in B. Graham (ed.), *In Search of Ireland*, London: Routledge, pp. 151–173.

Douglas, N. and Shirlow, P. (1998) 'People in conflict in place: the case of Northern Ireland', *Political Geography*, 17 (2): 125–128.

Downey, D. (2000) *The Guilt of Today*. Athlone: Red Line Books.

Du Gay, P. (ed.) (1997) *Production of Culture/ Cultures of Production*. London: Sage and Open University.

Dumper, M. (1996) *The Politics of Jerusalem since 1967*. Columbia: Columbia University Press.

Dunn, S. and Morgan, V. (1994) *Protestant Alienation in Northern Ireland: A Preliminary Survey*. Coleraine: Centre for the Study of Conflict.

Durkheim, E. (1915) *The Elementary Forms of the Religious Life*. London: George Allen & Unwin.

Eagleton, T. (1999) 'Nationalism and the case of Ireland', *New Left Review*, 234: 44–61.

Elliott, M. (2002) 'Religion and Identity in Northern Ireland', in M. Elliott (ed.), *The Long Road to Peace in Northern Ireland*. Liverpool: Liverpool University Press.

Ellis, G. (2001) 'Social exclusion, equality and the Good Friday Agreement: the implications for land use planning', *Policy and Politics*, 29 (4): 393–411.

Evans, M. and Syrett, S. (2003) *Generating Social Capital? The Social Economy and Local Regeneration*. Enfield: Middlesex University Press.

Feldman, A. (1991) *Formations of Violence: The Narrative of the Body and Political Terror in Northern Ireland*. Chicago: University of Chicago Press.

Fine, B. (2001) *Social Capital Versus Social Theory: The Political Economy and Social Science at the turn of the Millennium*. London: Routledge.

Foucault, M. (1979) *Discipline and Punish: The Birth of the Prison*. New York: Vintage.

Foucault, M. (1980) *Power/Knowledge: Selected Interviews and Other Writings 1972–1977*. New York: Pantheon.

Fraser, N. (1995) 'From redistribution to recognition? Dilemmas of justice in a post-socialist age', *New Left Review*, 212: 68–93.

Freeman, L. and Hamilton, D. (2004) 'The changing determinants of interracial home ownership disparities: New York City in the 1990s', *Housing Studies*, 19 (3): 301–332.

Gaffikin, F. and Morrisey, M. (2001) 'The other crises – restoring competitiveness to Northern Ireland's regional economy', *Local Economy*, 16 (1): 26–37.

Gallagher, A. (1994) 'Political Discourse in a Divided Society', in A. Guelke (ed.), *New Perspectives on the Northern Ireland Conflict*. Aldershot: Avebury.

Gallagher, M. (1995) 'How many nations are there in Ireland?', *Ethnic and Racial Studies*, 18 (4).

Gellner, E. (1986) *Culture, Identity and Politics*. Cambridge: Cambridge University Press.

Giddens, A. (1991) *Modernity and Self-Identity: Self and Society in the Late Modern Age*. Cambridge: Polity Press.

Gold, J. and Reville, G. (2003) 'Exploring landscapes of fear: marginality, spectacle and surveillance', *Capital and Class*, 80 (3): 27–50.

Gottman, J. (1973) *The Significance of Territory*. Charlottesville: University Press of Virginia.

Graham, B. (1997) 'The Imagining of Place: Representation and Identity in Contemporary Ireland', in B. Graham (ed.), *In Search of Ireland*. London: Routledge.

Graham, B. (1998) 'Contested images of place amongst Protestants in Northern Ireland', *Political Geography*, 17: 129–144.

Graham, B. and Shirlow, P. (1998) 'An elusive agenda: the development of the middle ground in Northern Ireland', *Area*, 30 (3): 245–254.

Graham, B. and Shirlow, P. (2002) 'The Battle of the Somme in Ulster memory and identity', *Political Geography*, 21 (7): 881–904.

Gramsci, A. (1971) *Selections from the Prison Notebooks*. London: Lawrence & Wishart.

Granovetter, M. (1973) 'The strength of weak ties', *American Journal of Sociology*, 78: 1360–1380.

Green, A., Shuttleworth, I. and Lavery, S. (2005) 'Young people, job search and labour markets: the example of Belfast', *Urban Studies*, 42 (2): 301–324.

Guelke, A. (2005) 'The Global Context: The International Systems and the Northern Ireland Peace Process', in J. Coakley, B. Laffan and J. Todd, *Renovation or Revolution? New Territorial Politics in Ireland and the United Kingdom*. Dublin: University College Dublin Press, pp.185–199.

Harbison, J. (2000) Latest FEC report (http://www.equalityni.org/whatsnew/archiveindiv.cfmStoryID=74).

Hargie, O. and Dickson, D. (2003) *Researching the Troubles*. Edinburgh: Mainstream Publishing.

Harvey, D. (2000) *Spaces of Hope*. Berkeley: University of California Press.

Healey, P. (2002) 'On creating the "city" as a collective resource', *Urban Studies*, 39 (10): 1777–1792.

Healey, P. (2004) 'The treatment of space and place in the new strategic spatial planning in Europe', *International Journal of Urban and Regional Research*, 28 (1): 45–67.

Heikkila, E. (2001) 'Identity and inequality: race and space in planning', *Planning Theory and Practice*, 2 (3): 261–275.

Hewitt, J. (1974) 'The Coasters', in P. Fiacc (ed.), *The Wearing of the Black: An Anthology of Ulster Poetry*. Belfast: Blackstaff Press.

Hodgson, G. (1988) *Economics and Institutions*. Cambridge: Polity Press.

Hoggart, K., Lees, L. and Davies, A. (2002) *Researching Human Geography*. Oxford: Arnold.

Home Office (2001) *Building Cohesive Communities: A Report of the Ministerial Group on Public Order and Community Cohesion*. London: Home Office.

Horowitz, D. (1986) *Ethnic Groups in Conflict*. Berkeley: University of California Press.

House of Commons (1996) *The Planning System in Northern Ireland. First Report of the Northern Ireland Affairs Committee*. London: HMSO.

Imrie, R. (2004) 'Urban geography, relevance, and resistance to the "policy turn"', *Urban Geography*, 25 (8): 697–708.

Jabri, V. (1998) 'Restyling the subject of responsibility in international relations', *Millennium*, 27 (3): 213–242.

Jackson, P. (1989) *Maps of Meaning: An Introduction to Cultural Geography*. London: Unwin Hyman.

Jarman, N. (1997) *Material Conflicts Parades and Visual Displays in Northern Ireland*. Oxford: Berg.

Jarman, N. (2005) *No Longer a Problem? Sectarian Violence in Northern Ireland*. Belfast: ICR.

Jarman, N. and O'Halloran, C. (2001) 'Recreational rioting: young people, interface areas and violence', *Child Care in Practice*, 7 (1): 2–16.

Jenkins, R. (1986) 'Northern Ireland: In What Sense "Religions in Conflict?"', in R. Jenkins, H. Donnan and G. McFarlane (eds), *The Sectarian Divide in Northern Ireland Today*. London: Royal Anthropological Institute of Great Britain and Ireland.

Keat, R. (2000) *Cultural Goods and the Limits of the Market*. Basingstoke: Macmillan.

Kennedy, D. (1991) 'Repartition', in J. McGarry and B. O'Leary (eds), *The Future of Northern Ireland*. Oxford: Clarendon Press.

Killias, M. and Clerici, C. (2000) 'Different measures of vulnerability in their relation to different dimensions of fear of crime', *The British Journal of Criminology*, 40: 437–450.

Knox, C. (1995) *Alienation: An Emerging Protestant Phenomenon in Northern Ireland*. University of Ulster: School of Public Policy, Economics and Law.

Knox, C. and Hughes, J. (1995) 'Local Government and Community Relations', in S. Dunn (ed.), *Facets of the Conflict in Northern Ireland*. Basingstoke: Macmillan.

KPMG (1995) *Interim Evaluation of the Gasworks Development: Laganside*. Belfast: KPMG.

LDC (Laganside Development Corporation) (2003) *Laganside Community Strategy*. Belfast: Laganside Development Corporation.

Lefebvre, H. (1991) *The Production of Space*. Oxford: Blackwell.

Lefebvre, H. (1995) *Writing on Cities*. Oxford: Basil Blackwell.

Levitas, R. (2000) 'Community, utopia and New Labour', *Local Economy*, 15: 188–197.

Lewis, C. (1999) 'The religiosity–psychoticism relationship and the two factors of social desirability', *Mental Health, Religion and Culture*, 3 (1): 39–45.

Ley, D. (1994) 'Gentrification and the politics of the new middle class', *Environment and Planning D: Society and Space*, 12: 53–74.

Lijphart, A. (1999) *Patterns of Democracy*. Boston: Yale University Press.

Lloyd, C., Shuttleworth, I. and McNair, D. (2004) *Measuring Local Segregation in Northern Ireland*. Belfast: School of Geography, QUB.

Maguire, S. and Shirlow, P. (2004) 'Shaping childhood risk in post-conflict Northern Ireland', *Children's Geographies*, 2 (1): 69–82.

Mandaville, P. (1999) 'Territory and translocality: discrepant idioms of political identity', *Millennium*, 28: 653–673.

Marcusse, P. (1993) 'What's so new about divided cities?', *International Journal of Urban and Regional Studies*, 17 (3): 355–365.

Massumi, B. (1993) 'Everywhere You Want to Be: Introduction to Fear', in B. Massumi (ed.), *The Politics of Everyday Fear*. Minneapolis: University of Minnesota Press.

McCann, E. (2005) *Rooting for England* (www.nuzhound.com/ articles/arts2005/ sep11_Northern_Ireland_acceptance__EMcCann_Sunday-Journal.php).

McCrudden, C., Ford, R. and Heath, A. (2004) 'The Impact of Affirmative Action', in R. Osborne and I. Shuttleworth (eds), *Fair Employment in Northern Ireland A Generation On*. Belfast: Blackstaff Press.

McGarry, J. and O'Leary, B. (1999) Policing *Northern Ireland: Proposals for a New Start*. Belfast: Blackstaff Press,.

McPeake, J. (1998) 'Religion and residential search behaviour in the Belfast Urban Area', *Housing Studies*, 13 (4): 527–548.

McVeigh, R. and Fisher, C. (2002) *Chill Factor or Kill Factor?* Belfast: West Belfast Economic Forum.

Melaugh, M. (1994) *Housing and Religion in Northern Ireland*. Coleraine: University of Ulster.

Michie, J. and Sheehan, M. (1998) 'The political economy of a divided Ireland', *Cambridge Journal of Economics*, 22 (2): 243–259.

Miller, R. (2004) 'Social Mobility in Northern Ireland: Patterns by Religion and Gender', in R. Osborne and I. Shuttleworth (eds), *Fair Employment in Northern Ireland A Generation On*. Belfast: Blackstaff Press.

Milton, K. (1993) 'Belfast: Whose City?', in C. Curtin, H. Donnan and T. Wilson (eds), *Irish Urban Cultures*. Belfast: Institute of Irish Studies.

Mitchell, P. (1999) 'Futures', in P. Mitchell and R. Wilford (eds), *Politics in Northern Ireland*. Boulder: Westview Press.

Moody, J. (2001) 'Race, school integration and friendship segregation in America', *American Journal of Sociology*, 107 (3): 679–716.

Morrow, D. (1996) 'In Search of Common Ground', in A. Aughey and D. Morrow (eds), *Northern Ireland Politics*. London: Longman.

Mouw, T. (2002) 'Are Black workers missing the connection? The effects of spatial distance and employee referrals on inter-firm racial segregation', *Demography*, 39 (3): 507–528.

Murray, M. and Murtagh, B. (2004) *Equity, Diversity and Interdependence*. Aldershot: Ashgate.

Murtagh, B. (1999) 'Listening to communities: locality research and planning', *Urban Studies*, 36 (7): 1185–1199.

Murtagh, B. (2002) *The Politics of Territory*. London: Palgrave.

Murtagh, B. (2003a) 'Territoriality, Research and Policy Making in Northern Ireland', in O. Hargie and D. Dickson (eds), *Researching the Troubles*. Edinburgh: Mainstream Publishing.

Murtagh, B. (2003b) *The Housing Executive Response to a Shared Future*. Belfast: Northern Ireland Housing Executive.

Murtagh, B. (2004) 'Collaboration, equality and land use planning', *Planning, Theory and Practice*, 5 (4): 453–469.

Murtagh, B. and McKay, S. (2003) 'Evaluating the social effects of the EU Community Initiative Programme', *European Planning Studies*, 11 (2): 194–211.

Myers, S. and Chung, C. (1998) 'Criminal perceptions and violent criminal victimization', *Contemporary Economic Policy*, 16: 321–333.

Nairn, T. (2001) 'Farewell Britannia: break-up or new union?', *New Left Review*, 7: 55–74.

Neill, W. (1995) 'Lipstick on the Gorilla? Conflict Management, Urban Development and Image Makers in Belfast', in W. Neill, D. Fitzsimmons and B. Murtagh (eds), *Reimaging the Pariah City*. Aldershot: Ashgate.

Neill, W. (1999) 'Whose city? Can a place vision for Belfast avoid the issue of identity?', *European Planning Studies*, 7 (3): 269–281.

Neill, W. (2004) *Urban Policy and Cultural Identity*. London: Routledge.

Nesbitt, D. (2005) *We deserve the facts about employment discrimination* (www. uup.org/dermotnesbitt_articles/07).

New Targeting Social Need (NTSN) Unit (1999) *Vision into Practice: The First New TSN Annual Report*. Belfast: Corporate Document Service.

Newman, D. (1999) *Territory, Boundaries and Postmodernity*. London: Frank Cass.

Northern Ireland Economic Research Council (NIERC) (2001) *Outlook Report*. Belfast: Queen's University.

Northern Ireland Housing Executive (NIHE) (1994) *Integration and Division, Housing Policy Review Discussion Paper*. Belfast: NIHE.

Northern Ireland Housing Executive (NIHE) (1998) *The Northern Ireland House Condition Survey 1996*. Belfast: NIHE.

Northern Ireland Housing Executive (NIHE) (2005) *Building Good Relations: Community Cohesion*. Belfast: NIHE.

Northern Ireland Housing Executive and the Eastern Health and Social Services Board (NIHE and EHSSB) (1995) *Health and Housing*. Belfast: NIHE.

Northern Ireland Office (1994) *Policy Appraisal and Fair Treatment Guidelines*. London: NIO.

O'Connor, F. (1993) *In Search of a State. Catholics in Northern Ireland*. Belfast: Blackstaff Press.

OECD (2000) *Urban Renaissance: Belfast's Lessons for Policy and Partnership*. Paris: OECD.

Office of the Deputy Prime Minister (ODPM) (2003a) *Cities, Regions, and Competitiveness*. London: ODPM.

Office of the Deputy Prime Minister (ODPM) (2003b) *Sustainable Communities: An Urban Development Corporation for the London Thames Gateway: A Consultation Paper*. London: ODPM.

Office of the Deputy Prime Minister (ODPM) (2003c) *Sustainable Communities: Communities Plan*. London: ODPM.

Office of the First and Deputy First Minister (OFMDFM) (2004) *An Anti-Poverty Strategy for Northern Ireland*. Belfast: OFMDFM.

Office of the First and Deputy First Minister (OFMDFM) (2005) *A Shared Future: Policy and Strategic Framework for Good Relations in Northern Ireland*. Belfast: OFMDFM.

O'Leary, J. and McGarry, J. (1993) *The Politics of Antagonism: Understanding Northern Ireland*. London: Athlone.

O'Leary, B. (1999) 'The nature of the British–Irish Agreement', *New Left Review*, 233: 66–92.

O'Leary, B. (2001) 'Election, not suspensions', *Guardian*, 13 July, p. 18.

Oliver, T. and White, O. (2004) *Ulster talks on track despite Paisley remark* (http://www.guardian.co.uk/Northern_Ireland/Story/0,2763,1363308,00.html).

Osborne R. (1996) 'Policy dilemmas in Belfast', *Journal of Social Policy*, 25 (2): 181–199.

Osborne, R. (2003) 'Progressing the equality agenda in Northern Ireland', *Journal of Social Policy*, 32 (3): 339–360.

Osborne, R. and Shuttleworth, I. (2004) 'Fair employment in Northern Ireland', in R. Osborne and I. Shuttleworth (eds), *Fair Employment in Northern Ireland A Generation On*. Belfast: Blackstaff Press.

Paasi, A. (2000) 'Territorial identities as social constructs', *Hagar*, 1: 93–95.

Paris C., Adair A., Berry, J. McGreal, S., Murtagh, B. and O'Hanlon, C. (1997) *Local Housing Market Analysis and Perceptions of Local Housing Markets within the Craigavon District Council Area. Final Report to the Northern Ireland Housing Executive*. Belfast: NIHE.

Parker, S. (2001) 'Community, social identity and the structuration of power in the contemporary European city, part one: towards a theory of urban structuration', *City*, 5 (2): 189–202.

Parkinson, M. (2004) *A Competitive City*. Belfast: Belfast City Council.

Pestieau, K. and Wallace, M. (2003) 'Challenges and opportunities for planning in the ethno-culturally diverse city: a collection of papers – Introduction', *Planning Theory and Practice*, 4 (3): 253–258.

Ploger, J. (2001) 'Public participation and the art of governance', *Environment and Planning, B* 28 (2): 219–241.

Poole, M. (1997) 'In Search of Ethnicity in Ireland', in B. Graham (ed.), *In Search of Ireland*. London: Routledge.

Poole, M. and Doherty, P. (1996) *Ethnic Residential Segregation in Belfast*. Coleraine: University of Ulster.

Porter, N. (1996) *Rethinking Unionism: An Alternative Vision for Northern Ireland*. Belfast: Blackstaff Press.

Porter, N. (2003) *The Elusive Quest: Reconciliation in Northern Ireland*. Belfast: Blackstaff Press.

Pratt, J. (2000) 'Commentary: economic change, ethnic relations and the disintegration of Yugoslavia', *Regional Studies*, 34 (8): 769–775.

Project Team (2002) *North Belfast Community Action Team*. Belfast: Project Team.

Putnam, R. (1983) *Making Democracy Work: Civic Traditions in Modern Italy*. Princeton: Princeton University Press.

Putnam, R. (1995) 'Bowling alone: America's declining social capital', *Journal of Democracy*, 6 (1): 65–78.

Putnam, R. (2000) *Bowling Alone: The Collapse and Revival of American Community*. New York: Simon & Schuster.

Raco, M. (2000) 'Assessing community participation in local economic development – lessons for the new urban policy', *Political Geography*, 19: 573–599.

Ray, L. and Sayer, A. (eds) (1999) *Culture and Economy after the Cultural Turn*. London: Sage.

Reskin, B., McBrier, D. and Kmec, J. (1999) 'The determinants and consequences of workplace sex and race composition', *Annual Review of Sociology*, 25 (4): 335–361.

Roger Tym and Partners (2003) *A Study to Establish Urban Development Corporation Boundaries in Thurrock*. London: ODPM.

Ruane, J. and Todd, J. (2003) 'Northern Ireland: Religion, Ethnic Conflict and Territoriality', in J. Coakley, *The Territorial Management of Ethnic Conflict*. London: Frank Cass.

Russell, R. (2004) 'Employment Profile of Protestants and Catholics', in R. Osborne and I. Shuttleworth (eds), *Fair Employment in Northern Ireland: A Generation On*. Belfast: Blackstaff Press.

Sack, R. (1986) *Human Territoriality: Its Theory and History*. Cambridge: Cambridge University Press.

Sack, R. (1997) *Homo Geographicus: A Framework for Action, Awareness and Moral Concern*. Baltimore: Johns Hopkins University Press.

Sack, R. (1999) 'A sketch of a geographical theory of morality', *Annals of the Association of American Geographers*, 89 (1): 26–44.

Sack, R. (2003) *A Geographical Guide to the Real and the Good*. New York, London: Routledge.

Saunders, D. (1990) *A Nation of Home Owners*. London: Unwin Hyman.

Saunders, P. (1986) *Social Theory and the Urban Question*. London: Hutchison.

Sayer, A. (2000) 'System, lifeworld and gender: associational versus counterfactual thinking', *Sociology*, 34 (3): 707–725.

Sheehan, M. and Tomlinson, M. (1996) 'Long-term Unemployment in West Belfast', in E. McLaughlin and P. Quirk (eds), *Policy Aspects of Employment Equality in Northern Ireland*. Belfast: SACHR.

Shirlow, P. (1997) 'Class, Materialism and the Fracturing of Traditional Alignments', in B. Graham (ed.), *In Search of Ireland*. London: Routledge, pp. 87–107.

Shirlow, P. (2001) 'Devolution in Northern Ireland/ Ulster/ the North/ six counties: delete as appropriate', *Regional Studies*, 35 (8): 743–752.

Shirlow, P. (2003a) 'Ethno-sectarianism and the reproduction of fear in Belfast', *Capital and Class*, 80 (2): 77–94.

Shirlow, P. (2003b) '"Who Fears to Speak": fear, mobility and ethnosectarianism in the two Ardoynes', *Global Review of Ethnopolitics*, 3 (1): 76–92.

Shirlow, P. and McGovern, M. (1997) *Who Are 'The People'? Protestantism, Unionism and Loyalism in Northern Ireland*. London: Pluto Press.

Shirlow, P. and McGovern, M. (1998) 'Language, discourse and dialogue: Sinn Fein and the Irish peace process', *Political Geography*, 69 (2): 171–186.

Shirlow, P. and Monaghan, R. (2004) 'Northern Ireland: ten years on', *Terrorism and Political Violence*, 15 (3): 397–400.

Shirlow, P. and Murtagh, B. (2004) 'Capacity building, representation and intra community conflict', *Urban Studies*, 41 (1): 57–70.

Shirlow, P. and Pain, R. (2003) 'Introduction: the geographies and politics of fear', *Capital and Class*, 80: 15–26.

Shirlow, P. and Shuttleworth, I. (1999) 'Who is going to toss the burgers', *Capital and Class*, 69 (2): 27–46.

Shirlow, P. and Stewart, P. (1999) 'Northern Ireland: between peace and war?', *Capital and Class*, 69 (2): 1–12.

Shirlow, P., Murtagh, B., Mesev, V. and McMullan, A. (2003) *Measuring and Visualising Labour Market and Community Segregation*. Belfast: OFMDFM.

Shuttleworth, I. and Green, A. (2004) 'Labour Market Change in Northern Ireland: Unemployment, Employment and Policy', in B. Osborne and I. Shuttleworth (eds), *Fair Employment in Northern Ireland: A Generation On*. Belfast: Blackstaff Press.

Shuttleworth, I., Shirlow, P. and McKinstry, D. (1996) 'Vacancies, Access to Employment and the Unemployed: Two Case Studies of Belfast and (London)Derry', in E. McLaughlin and P. Quirk (eds), *Policy Aspects of Employment Equality in Northern Ireland*. Belfast: SACHR.

Sibley, D. (1995) *Geographies of Exclusion*. London: Routledge.

Simmel, G. (1955) *Conflict and the Web of Group-Affiliation*. New York: Free Press.

Singleton, D. (1986) *Aspects of Housing Policy in Northern Ireland*. Belfast: Queens University.

Smelser, N. and Swedberg, R. (eds) (1994) *The Handbook of Economic Sociology*. New York: Russell Sage Foundation.

Smith, A. (1991) *Nations and Nationalism in a Global Era*. Cambridge: Polity Press.

Smith, D. and Chambers, G. (1989) *Equality and Inequality in Northern Ireland, Report No. 4, Public Housing*. London: Policy Studies Institute.

Smith, M. and Alexander, K. (2001) 'Building a Shared Future: The Laganside Initiative in Belfast', in W. Neill and H. Schwedler (eds), *Urban Planning and Cultural Inclusion*. London: Palgrave.

Soja, E. (1989) *Postmodern Geographies. The Reassertion of Space in Critical Social Theory*. London: Verso.

Soja, E. (1999) 'In different spaces: the cultural turn in urban and regional political economy', *European Planning Studies*, 7 (1): 65–75.

Soja, E. (2000) *Postmetropolis: Critical Studies of Cities and Regions*. London: Blackwell.

Sorensen, B. (2003) *The Organisational Demography of Racial Employment Segregation.* (http://web.mit.edu/~sorensen/www/workplace%20segregation.Pdf).

Squires, G. and Kubrin, C. (2005) 'Privileged places: race, uneven development and the geography of opportunity in urban America', *Urban Studies*, 42 (1): 47–68.

Standing Advisory Commission on Human Rights (SACHR) (1990) *Religious and Political Discrimination and Equality of Opportunity in Northern Ireland, Second Report.* London: HMSO.

Stewart, P. and Shirlow, P. (1999) 'Northern Ireland between peace and war', *Capital and Class*, 69: 1–12.

Storey, D. (2001) *Territory: The Claiming of Space.* Harlow: Prentice Hall.

Taylor, P. (1994) 'The State as container: territoriality in the modern world-system', *Progress in Human Geography*, 19: 1–15.

Taylor, R. (1988) *Human Territorial Functioning.* Cambridge: Cambridge University Press.

Tewdwr-Jones, M. (2003) 'Urban development, civic planning and the Janus Syndrome', *International Planning Studies*, 8 (4): 251–252.

Throgmorton, J. (2004) 'Where was the wall then? Where is it now?', *Planning Theory and Practice*, 5 (3): 349–365.

Todd, J. (2005) 'Northern Ireland: The Changing Structure of Conflict', in J. Coakley, B. Laffan and J. Todd, *Renovation or Revolution? New Territorial Politics in Ireland and the United Kingdom.* Dublin: University College Dublin Press.

Tonge, J. (2004) *The New Northern Irish Politics?* Basingstoke: Palgrave Macmillan.

Tulloch, J. (2004) 'Parental fear of crime: a discursive analysis', *Journal of Sociology*, 40: 362–377.

Tulloch, J. and Lupton, D. (2003) *Risk and Everyday Life.* London: Sage.

Tyler, P. (2004) *Turning Belfast Around.* Belfast: Belfast City Council.

Wasserstein, B. (2001) *Divided Jerusalem.* London: Profile Books.

Weiner, R. (1980) *The Rape and Plunder of the Shankill.* Belfast: Farset Press.

Whyte, J. (1990) *Interpreting Northern Ireland.* Oxford: Clarendon Press.

Wilford, R. and Wilson, R. (2001) *A Route to Stability? The Review of the Belfast Agreement.* Belfast: Democratic Dialogue.

Wilford, R. and Wilson, R. (2003) *Northern Ireland: A Route to Stability.* Swindon: ESRC.

Williams, P. and Chrisman, L. (1993) *Colonial Discourse and Post-Colonial Theory. A Reader.* New York: Harvester Wheatsheaf.

Wilson, R. (2003) *Northern Ireland: What's Going Wrong?* London: Constitutional Unit (UCL).

Wilson, T. (1989) *Ulster: Conflict and Consent.* Oxford: Basil Blackwell.

Wilson, T. and Donnan, H. (1998) *Border Identities: Nation and State at International Frontiers.* Cambridge: Cambridge University Press.

Index

Compiled by Sue Carlton